LIFE SCIENCE LIBRARY

TIME

TIME LIFE BOOKS ®

This volume is one of a series that explores the world of science, from the crude observations of the first astronomers to 21st Century technology.

TIME

by Samuel A. Goudsmit, Robert Claiborne
and the Editors of **TIME-LIFE BOOKS**

Revised Edition

TIME-LIFE BOOKS   ALEXANDRIA, VIRGINIA

CONTENTS

ABOUT THIS BOOK

WHAT IS TIME? Any child knows the answer, and yet even the most advanced theoretical physicist is hard put for a fully satisfactory definition. He cannot say for certain when time started, when it will end, or even if it really exists in a philosophical sense. Yet the measurement of time is the basis of all science, for the scientist can study only what changes with time. Astronomers chart the history of the universe in "big time," the passage of billions of years. Physicists and engineers subdivide "little time" into billionths of a second. Biologists have discovered that animals and plants measure time, too—that even the lowliest single-celled organisms depend on biological clocks to keep themselves synchronized internally and externally. This book explores all of these facets of the meaning of time.

Each chapter is followed by a supplementary picture essay which may be read independently. For instance, Chapter 4, which explores the history of man's search for more accurate clocks, is followed by an essay which explains how several of the earliest clocks worked.

THE AUTHORS

SAMUEL A. GOUDSMIT of the University of Nevada is a former editor-in-chief for the American Physical Society and formerly a physicist at Brookhaven National Laboratory. His discovery in 1925 that electrons spin on their axes helped establish modern atomic theory. He wrote the book *Alsos* about his World War II role in scientific intelligence. His awards include the Medal of Freedom and the Order of the British Empire.

ROBERT CLAIBORNE, a former chemistry student who switched to journalism, specializes in writing and editing scientific material. He was Associate Editor for *Scientific American* and later Managing Editor of *Medical World News*. He is a former Editor of the LIFE Science Library and author of *Climate, Man and History* and *On Every Side the Sea*. He has contributed articles to *The Nation*, *Harper's* and other leading magazines.

ON THE COVER

A 19th Century pocket watch offers an elegant example of man's interest in time.

1

The Elusive
Nature
of Time

American Standard Time has been determined since 1865 by more than 70 master
clocks at the U.S. Naval Observatory. This
master clock, the Observatory's 64th,
served for three years before its millionth-
of-a-second accuracy became obsolete.

THROUGHOUT ALL OF MAN'S EXPERIENCE, through every aspect of the world and universe he inhabits, runs the elusive entity called time. The clock, deputy for the sun and stars, tells him it is time to get up, time to go to school or to work, time to put the roast in the oven or to eat it, time to retire. Setting out on a journey, he checks timetables of train or plane against his watch; putting out to sea, he must correlate time and distance to find his way.

Time governs not merely man's activities but his very being. Like every living organism, he exists by grace of thousands of intricately synchronized rhythms. His pulse keeps time, tranquilly or otherwise; the electrical waves in his brain time their rhythms to sleep or wakefulness.

Other living creatures, far more than man, are governed by "biological time" that links interior processes to the regular rhythms of the outside world. The morning glory opens by the clock; the maple leaf grows green or flames scarlet by the calendar; the mallard flies north or south—taking its direction from some instinctive internal calculus involving time and the sun.

Time, which gives continuity and pattern to life, also brings disruption and death. The morning glory, splendid in the dawn, is wilted at noon; man, maple and mallard live their allotted span. The eternal hills were not there a hundred million years ago and will be gone a hundred million years hence; even the stars shift with the centuries and, early or late, must eventually fade out. There is nothing under the sun, or over it, of which we cannot say: "This, too—in time—will pass away."

Of all the great abstractions of science, it is omnipresent time—not space or force or matter—that comes most often to our lips. Time is a great teacher, a great healer, a great legalizer and leveler; it stands still, slips away from us or flies past us. We can save time or lose it, spend time or waste it (time is money!), even beat it or kill it.

What we cannot do, oddly enough, is define it. To the psychologist, time is an aspect of consciousness, the means by which we give order to our experiences. To the physicist time is one of the three fundamental quantities—the other two are mass and distance—in terms of which he can describe anything in the universe. To the philosopher time is still other things. Yet these learned men, though they may write books about time, can none of them define it in a way satisfactory to one another, or even to themselves.

The puzzle and paradox of indefinable time was summed up 1,500 years ago by Aurelius Augustinus, bishop of Hippo in North Africa, philosopher and later saint. "What then is time?" he asked. "If someone asks me, I know. If I wish to explain it to someone who asks, I know not." Fifteen centuries have not sufficed to solve St. Augustine's problem.

Thinking and talking about this indefinable entity is made no easier by the fact that the word "time," scientifically speaking, refers to two different, though related, things. The first is *interval*, which means duration in time. The second is *epoch*, which means location in time. If one asks, "How long will the concert last?" one is talking about interval. If one asks, "What time will the concert begin?" the subject is epoch. We make a similar distinction, in talking about space, between length and location: "My house is 30 feet (9.1 m) wide" versus "My house is 30 feet from the corner." Both interval and epoch are expressed in the same units—days, hours, minutes and so on—but are not the same thing. And the difference between them is often important but rarely obvious.

The unique human time sense

Man's puzzlement and preoccupation with time both derive ultimately from his unique relationship to it. All animals exist in time and are changed by it; only man can manipulate it.

Like Proust, the French author whose experiences became his literary capital, man can recapture the past. He can also summon up things to come, displaying imagination and foresight along with memory. It can be argued, indeed, that memory and foresightedness are the essence of intelligence; that man's ability to manipulate time, to employ both past and future as guides to present action, is what makes him human.

To be sure, many animals can react to time after a fashion. A rat can learn to press a lever that will, after a delay of some 25 seconds, reward it with a bit of food. But if the delay stretches beyond 30 seconds, the animal is stumped. It can no longer associate reward so "far" in the future with present lever-pressing.

Monkeys, more intelligent than rats, are better able to deal with time. If one of them is allowed to see food being hidden under one of two cups, it can pick out the right cup even after 90 seconds have passed. But after that 90-second time interval, the monkey's hunt for the food is no better than chance predicts.

With the apes, man's nearest cousins, "time sense" takes a big step forward. Even under laboratory conditions, quite different from those they encounter in the wild, apes sometimes show remarkable ability to manipulate the present to obtain a future goal. A chimpanzee, for example, can learn to stack four boxes, one atop the other, as a platform from which it can reach a hanging banana. Chimpanzees, indeed, carry their ability to cope with the future to the threshold of human capacity: They can make tools. And it is by the making of tools—physical tools as crude as a stone chopper, mental tools as subtle as a mathematical equation—that man characteristically prepares for future contingencies.

Chimpanzees in the wild have been seen to strip a twig of its leaves to make a probe for extracting termites from their hole. Significantly, however, the ape does not make this tool *before* setting out on a termite hunt, but only when it actually sees the insects or their nest. Here, as with the banana and the crates, the ape can deal only with a future that is immediate and visible—and thus halfway into the present.

When the small-brained ape-man *Australopithecus* began to make tools, some two million years ago, his first crude stone choppers, like the chimpanzee's stripped twig, may have been improvised to meet an immediate and visible future. Before long, however, he evolved to the

EGYPT'S GOD OF TIME, ibis-headed Thoth *(above),* was said to have reckoned the divisions of time in the Egyptian calendar, and the first month was named for him. He also was supposed to weigh the souls of the dead to determine which would live on in the timeless afterworld. Here he is shown with the symbol of life in his left hand and a stylus to record the fate of the dead in his right hand.

point where he began carrying his improvisations about with him. A piece of shaped quartz, unearthed several miles from the nearest quartz deposit, shows that its maker possessed at least enough foresight to hold on to a rock that he had taken the trouble to batter into a useful shape.

With bigger brains came greater foresight. Perhaps 500,000 years ago, the first men in China used fire. And the regular use of fire, as the anthropologist F. Clark Howell says, implies men "provident enough to keep supplies of fuel on hand and skillful enough to keep fires going."

Certainly by this time, and probably earlier, man had developed his most remarkable tool, and one that revolutionized his relationship to time: language. Words are not simply means of communication; many beasts and birds communicate. Words are the best tools for "moving" things in time. To name a thing or an action is to call it to mind, to summon it up, from past or future. (Even today, to "speak of the Devil" is to invite his presence!) With the aid of words, man could mull over past time and plan for future time with new precision. (Try deciding what you are going to have for dinner without naming things.) He could instruct his sons at leisure on how to meet the charge of some future mammoth or wild ox, instead of having to reserve his gestured or grunted warnings for the moment of peril.

The development of true foresight

Neanderthal man, with a brain at least as big as our own, raised foresight and imagination to near-modern levels. He made not only tools for future needs but also tools to make future tools: blades for cutting and whittling wood; notched pieces of stone presumably used for shaping spear tips; chisels and scrapers. He seems also to have practiced some sort of hunting magic, which is a tool of the imagination. Finally, Neanderthal man buried his dead and interred with the bodies tools that argue persuasively for some sort of belief in an afterlife. For the Neanderthals, as for ourselves, the future must have stretched not merely far into the known but into the unknown as well.

Man's innate capacity for dealing with time does not seem to have increased markedly since the Neanderthals walked the earth. By contrast, his cultural equipment for coping with it has evolved tremendously. It is fair to say that without a growing consciousness of time, the tools to measure time and the concepts to relate it to other things, civilization would be totally impossible.

The development of writing, considered the hallmark of civilization, is itself a landmark in the conquest of time. Man's memory, though it helped to make him human, remained humanly fallible. With writing, a sort of artificial memory, he could begin to keep accurate records and could even read, and sometimes profit by, the experiences and thoughts of men long dead.

With writing and record-keeping came techniques for measuring time. The calendar connected human activities more accurately with the seasons and also made possible the coordination of the efforts of many

FATHER TIME is believed to stem from Cronus, the Greek god of agriculture and father of Zeus. Later Greeks knew him as Chronos, god of time, and his sickle became an implement for cutting down the passing years. Today he symbolizes the end of the year. Born at the moment the year begins, he lives a complete life cycle within the year and then is reborn to begin the new year.

thousands of men separated by distance. Clocks, reckoning time not by months or days but by hours, minutes and seconds, made possible tighter synchronization of more complex societies. Ultimately, as clocks were devised that could measure billionths and trillionths of a second, they became built into the intricate technology that serves civilization.

Clocks that measure the earth's age

If time measurement has been the handmaiden of civilization, it has been the very companion and partner of science. Science made time measurement possible: Calendars and clocks alike are based on astronomical observations of the changing heavens; Galileo's studies of the pendulum opened a new chapter in clockmaking; the most precise modern timekeepers derive from properties of the atom identified by scientists during the past 50 years. Equally, time measurements have made science possible. Early clocks enabled Galileo and his successors to solve the mysteries of motion, thus laying the groundwork of modern physics. More accurate clocks enabled man to map the earth with precision and fix his own position upon it. Still more accurate clocks have helped to unravel some of the fundamental secrets of matter and energy, provided clues to changes deep within the earth and, indeed, have radically reshaped our concepts of time itself. And yet other clocks, which "tick" by centuries or millennia rather than by seconds, have enabled scientists to date the eras of Neanderthal and *Australopithecus* on earth and even to measure how long the earth itself has hung in the heavens.

But neither the measurement of time nor the civilization and science based on time measurement would be conceivable without man's awareness of time. How he achieves his concepts of time is therefore fundamental to an understanding of it. If the idea of past, present and future is uniquely human, how do these notions develop? If time is, in one sense, "how long we have to wait," how long is that—and how do we know?

A newborn baby, like most lower animals, lives in an eternal present. He knows no past: If his groping hands have seized some dangerous object, his mother need only distract him for an instant, hide the forbidden object, and for him it has never existed. He knows no future: If he is hungry, he cannot wait and his wails can be stopped only by food.

In a few weeks, he has moved up the evolutionary scale a bit. His hungry cries can now be stopped, not merely by the bottle itself but also by the mere sight of the bottle. Like the laboratory rat pressing a lever for a delayed reward, he is learning to associate present events with future consequences. By the time he has passed the age of two, he can emulate (and quickly surpass) the monkey and the ape in finding a hidden object after a delay. A child of two and a half who has seen a toy hidden in one of three boxes can pick out the right box nearly a minute later; when he is six, he can do so after a wait of 35 minutes.

By the time a child has acquired an adult ape's ability to manipulate things, he has already left the ape far behind, for he has learned to use language. Soon after the age of one and a half he will talk about objects

STONE AGE TIME SENSE is indicated by these tools, a Paleolithic hand ax *(left)* and a Neolithic hoe. Since they were made for later use, they imply that Stone Age man was aware of the future and planned for it. The 400,000-year-old ax is believed to be the first standardized tool made by man and was probably a general-purpose tool like a jackknife used for cutting and scraping. The hoe was a more sophisticated tool designed to till the ground. This one dates from about 5000 B.C., in the late Stone Age, when men were just beginning to organize an agricultural society.

and events that are *not* present; by three he will speak of what he wants to do next day. In a few more months he will begin to be aware of the day as a unit of time. ("When it gets to be dark time I have my supper and go to bed; when the dark is gone, I get up and have my cereal.") By four he will talk about "next summer" and begin to look forward to important coming events like birthdays and Christmas.

A growing feeling for sequence

But not for years can the child organize his time sense to realize that if Johnny is older than Paul, then he must have been born *before* Paul, not after. He may be aware of individual sequences, but not of the connections between them. Witness the following characteristic dialogue between a six-year-old and a psychologist:

"Is your daddy older or younger than you?"

"Older."

"Was he born before or after you?"

"Don't know."

"Who came first, he or you?"

"Me."

"Do you always stay the same age or do you grow older?"

"I get older."

"How about your father?"

"He always stays the same age."

By eight the child can relate sequences, and his concept of the future, like that of Neanderthal man, has expanded to include activities that he has never known himself: "When I grow up. . . ." Soon his time sense is fully developed. As an adult his memory of the past extends back to childhood; his plans for the future may stretch for weeks and even years ahead—camping equipment for a vacation six months off, investments for a retirement 20 years away.

A normally developed time sense is, indeed, one mark of a normal adult, for disturbances of time sense are often a symptom of mental abnormality. Many retarded individuals show difficulties in organizing their memories. They can recall the order of incidents during the two or three weeks just past, but occurrences prior to that period are all lumped together in a sort of "past indefinite." Like young children, they are "prisoners of the present."

More oddly perverse is the singular time sense of some idiots savants, such as the identical twins "George" and "Charles" described by an American psychologist. These remarkable youths, whose I.Q.s of 70 were far below average, could recall the most trivial incidents, including detailed weather conditions, on any named date for years back. They could answer almost instantly such questions as, "On what date will the first Wednesday in July fall in the year 2002?" Yet they could not tell how much change a 35-cent purchase leaves from a dollar bill!

An inability to cope with time is often a symptom of a serious mental illness such as schizophrenia. In the words of one patient, "I just can't

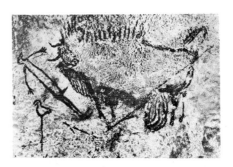

A CAVEMAN SEES THE FUTURE in this prehistoric wall painting in a Lascaux cave in France. It seems to portray the concept of a life after death—which implies a relatively advanced idea of future time. The painting, which was done some 25,000 years ago by a Cro-Magnon man, shows a wounded bison attacking and killing its hunter. A little bird, which is believed to represent the dying man's soul, sits nearby, apparently ready to fly away as soon as the body is dead.

seem to grasp the fact that time passes and the hands of the clock go round. Sometimes, outside in the garden when they run quickly up and down . . . or the leaves whirl in the wind, I wish I could live again as before and be able to *run with them* within me so that time would pass again. But there I stop and I do not care . . . I just bump into time."

The physiology of past-present-future

The psychology of past-present-future—how it develops, fails to develop or retrogresses—is partly understood. But its physiology—the neurological processes that make it work—is only now beginning to be understood. The brain's frontal lobes seem to have something to do with the time sense; this is predictable, since it is the high development of these areas in the brain that most sharply sets man apart from lower animals. Mental patients whose frontal lobes have been partially severed from the rest of the brain to relieve severe depression show a marked tendency to live in the present. Monkeys whose lobes have been removed show an even more contracted time horizon; under certain conditions, they appear to remember for no more than five seconds.

Yet lobotomized monkeys can still solve fairly complex problems involving time. For example, they can, with no special difficulty, learn to stay on one side of the cage for 10 seconds, then shift to the other side during the following 10 seconds. Scientists suspect that both monkey and man possess some other time sense, which resides outside the frontal lobes in the pineal gland, deep in the brain. It has been established that this sense concerns not the "before or after" aspects of time but its duration, that is, interval time alone.

Subjective time intervals, the "how long" in how long we have to wait, are notoriously hard to estimate. The busy housewife turns her back and the pot boils over; had she watched it, of course, it would never have boiled. French coal miners, trapped underground, emerge after three weeks of frantic digging—and remark that they have undergone a rough four or five days. If we are idle or bored, time drags; if we are occupied, time flies. "Work," said the philosopher Diderot, "has the advantage, among others, of shortening our days and lengthening our lives."

Despite these and many other oddities that warp our estimates of time interval, subjective time is by no means "all in the mind." Evidence both psychological and physiological points to the existence of some sort of bodily "pacemaker" which influences our perception of time, whether we are busy or bored.

A person under hypnosis can judge time—for instance, by awakening after a specified interval—more accurately than he can estimate it awake. Involuntary responses have been proved to measure intervals better than do conscious judgments. In one experiment several subjects were repeatedly stimulated by hearing a noise and then, 9.4 seconds later, seeing a flash of light. Eventually they developed a conditioned reflex, in which the buzzer by itself produced a delayed change in their brain waves. The brain "measured" time fairly accurately: The delay between noise and brain response ranged from 7.2 to 9.2 seconds. However, when the same subjects were asked to make a conscious estimate of the delay between sound and flash, their guesses were much wider of the mark, ranging from 6.0 to 15.2 seconds.

When time stretches

If the body contains some physiological rhythm by which the brain can unconsciously measure interval time, changes in physiology should make the rhythm faster or slower. Such, in fact, seems to be the case. One striking example of physiological alteration in subjective time was studied by the American biologist Hudson Hoagland, during an illness of his wife. Mrs. Hoagland, as the result of an infection, ran a fever of 103 (39.4° C.). Her husband went to a drugstore for medicine and returned 20 minutes later. His wife, however, insisted he had been gone much longer. Struck by her conviction, he asked her to count off 60 seconds to herself while he timed her with a stopwatch; he found that her counting had actually taken only 37.5 seconds. As her recovery progressed, he repeated the test several times, and as the fever dropped, her rate of counting slowed. Her estimate of a one-second interval was returning to normal. From this and other experiments, Hoagland concluded that the brain contains some sort of chemical pacemaker whose activity, like all chemical processes, can be speeded up by heat.

The pacemaker can also be thrown off by drugs. The English writer Thomas De Quincey reported that under the influence of opium he seemed to live as much as 100 years in a single night. Another Englishman, J. Redwood Anderson, reported that after taking hashish, "Time was so immensely lengthened that it practically ceased to exist." In general, stimulants like caffeine and amphetamine, which speed up metabolism, make time seem longer; depressants like the barbiturates or opiates make it seem shorter.

So easily disoriented, the brain's pacemaker gives man only a vague and untrustworthy instinct for time. In this innate ability he is far surpassed by the plants and animals, which, as the following chapter will describe, reliably count days and seasons with internal "clocks" synchronized to sun or stars. Yet man's instinctual shortcomings are more than made up for by his great powers of abstraction, which have enabled him to create clocks and calendars; to anticipate beyond the seasons and recapture the distant past; and to attempt, in ever more fruitful ways, to define the indefinable entity that is time itself.

A TEST OF SUBJECTIVE TIME reveals that people's concepts of the relationship of past events vary widely with their ages. In this test four subjects, ranging in age from nine to 70 years, were given strips of paper (below) which represented their life-spans from birth to the present. They were asked to indicate five points in time from "yesterday" back to when they entered the first grade. Although all of the subjects gave more weight to recent times—marking yesterday and last week farther to the left than they actually should be—time sense seemed to become more realistic with age. The nine-year-old marked yesterday, last week and six months ago at almost equal intervals, while the 70-year-old spaced them at roughly appropriate distances.

1. YESTERDAY
2. LAST WEEK
3. 6 MONTHS AGO
4. 5 YEARS AGO
5. 1ST GRADE

Years,
Months, Minutes

The hands of the clock merge at midnight. Suddenly the room explodes with excitement—for this is not just any midnight, but New Year's, the starting point for one of man's most important units of time. There are many such units: Some of them are natural (such as the year, the time the earth takes to make one circuit of the sun, and the day, the time the earth takes to revolve on its axis), but other units are man-made. For the need to chop time into manageable units is so powerful that when nature has not supplied the divisions, man has created them to suit his needs.

Thus he has strung together seven of nature's days to make a week, providing an appropriate period for various routines —marketing, rest, worship. He has subdivided the day into hours, to gauge the daily routines of waking, eating and labor. He has broken up the hour into minutes, minutes into seconds (though not until the 18th Century) and, in recent times, seconds into such smaller segments as milliseconds and microseconds, primarily useful in science and technology.

Natural or artificial, these divisions of time have become vital to man and his way of life. With them, he is able to control time. Without them, he would be at time's mercy.

STARTING THE YEAR ANEW
New Year's, which affords man the opportunity to start afresh, is greeted with jubilation. But here in a San Francisco nightclub, the holiday is celebrated every night—with good reason, for New Year's could fall on any day. January 1 is arbitrarily designated by society to mark the end of one revolution of the earth around the sun and the beginning of another.

Climaxing a once-in-a-lifetime pilgrimage to Mecca, the Moslem faithful "stand before God" from noon to sunset on the Mount of Mercy, where

Moments of a Lifetime

To any man, the most important single unit of time is that of his own stay on earth. He marks off his personal span in personal terms, by the major events that happen only once. His birth, his marriage, the arrival of his children make milestones which he commemorates with birthdays and anniversaries. His acceptance into adult society may be noted by confirmation, bar mitzvah, graduation from school or, in more primitive societies, by tests of strength and skill.

To many a man, some other single event marks the most significant moment of his life—a personal triumph, a tragedy, a journey, the revelation of a faith. To a devout Moslem, the goal of a lifetime is a pilgrimage to Mecca. Year after year, pilgrims descend on this holy city from every corner of the globe. First they visit the

Mohammed delivered his final sermon. Many carry umbrellas as protection from the sun, since pilgrims are forbidden to wear head coverings.

holiest shrine in Islam, the Kaaba, a cubical stone building, and circle it seven times, stopping on each circuit to kiss a holy meteorite housed inside; then comes the Lesser Pilgrimage to the plains between Safa and Marwa; on the ninth day is the Greater Pilgrimage to the Mount of Mercy where the pilgrims "stand before God" (above), fervently celebrating the spiritual high point of a lifetime.

Men Tied to the Seasons

The seasons have always governed men's lives and in turn have been adopted as basic units of time. To anticipate the season's arrival helps man gain his food, telling him when to hunt animals and when to plant his crops. Modern industry depends on seasonal movements of goods: clothes, cars, air conditioners, even toys—all are made and sold according to seasonal needs.

Some societies are dominated by seasons. Laplanders follow the seasonal migrations of the reindeer. The nomadic tribes of Iran and Afghanistan, shown here, uproot themselves from the valleys in the spring to seek the rich grasslands in the mountains. In the cold weather, they return to the valleys. To these people, shuttling between mountain and valley, time is broken in two by the seasons.

THE WAY OF WANDERERS
Tended by the Bakhtiari tribe in Iran *(left)*, flocks of sheep start a 75-mile trek in the spring for the green mountain pasturelands where they will spend the summer grazing. In Afghanistan *(above)*, two veiled women flank a young boy as they ride toward spring grasslands.

The Mystical, Practical Month

The month, like the seasons, was given to man by nature. Its origin rests in the length of time—about 29 days—it takes for the moon to orbit the earth. Since the moon has long been an object of wonder and veneration, the month has taken on almost mystical connotations in many societies. In Asia, for example, the moon governs some Buddhist religious observances, such as the one shown here.

But in the West the month is less strictly measured by the movements of the moon and has become a time unit of convenience, a useful way to subdivide the year and the seasons. One of the chief functions of the month is as a time unit of commerce and of the periodic exchange of money. Bills, rent and credit payments are usually due each month; for many people payday is a monthly event.

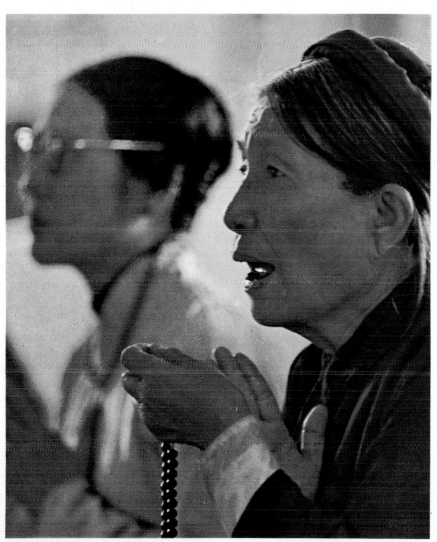

WORSHIPING BY THE MOON
Buddhist monks and laymen (left) meet for prayer and meditation at the Xa Loi pagoda in Ho Chi Minh City, Vietnam, twice during the lunar month—when the moon is new and when it is full. These ceremonies have been traced to ancient sects of moon worshipers; for the Buddhists they are times of spiritual rejuvenation. The faithful fast and attend temple services, where they are exhorted to think with clean minds and hearts. Above, two women at Xa Loi offer prayers; the one in the foreground is holding a 108-bead Buddhist rosary.

The Special Needs for a Week

Every society has had to invent the week, for a time unit smaller than a month but longer than a day is essential to human affairs. A community functions more smoothly if regularly recurring days are set aside for laundry, marketing, time off from the job and worship.

The ancient Greeks split their months into three 10-day weeks and the Romans had a market week of eight days ending with a day of rest and festivals. Among primitive tribes today, the market week varies in duration from four to 10 days. The seven-day week is practically unknown among these societies; it is the only week-length not in use by them.

Among the major nations of the world, however, the seven-day week is the custom. This length was derived from Genesis—"And on the seventh day, God rested"—and was established over most of the world by the spread of Christianity and Mohammedanism. Attempts to change the seven-day week in Western societies have invariably failed. In 1792 the French Revolutionary Convention enacted a decimal calendar that called for 10-day weeks. Although the calendar worked, it was abandoned after Napoleon came to power. In our own century, the U.S.S.R. tried twice to alter the week, decreeing a week of five days in 1929 and a week of six days in 1932. But by 1940, the seven-day week had been restored.

THE HOLIDAY OF MARKETING
In the bustling market place of Dubrovnik, Yugoslavia, merchants and buyers gather together, as they have for centuries, to barter, to buy and to exchange goods. Market day is traditionally a holiday and a time of festivals and parties as everyone relaxes after the week's labor.

Dawn finds a Swiss milkboy already at work. He arises at 5 a.m. to start the daily milking on a mountain farm above the village of Elm.

The Productive Cycle of the Day

Since man first walked the earth, the day has been the most obvious indicator of time's passage. Nothing exerts a greater control on man's sense of time and activity than the blaze of light and warmth that floods the world at dawning, and the cool darkness that occurs when the sun sets.

With each sunrise, man begins to work—milking his cows, selling his produce, toiling in factory or office. As the sun crosses the sky, the tempo of the day increases; as the sun begins to set, man's working hours come slowly to a halt. Farm animals are bedded down, shops are closed and it is time for family, food and relaxation. Soon the dark side of the earth is falling asleep, with only man's own artificial lights left to take the place of the departed sun—while half the globe away the sun is just rising and people are stirring as they make ready to begin their daily routine.

At sunset, a fish seller in Chicago's northwest section prepares to close his shop, his work finished with the end of the daylight hours.

The Hour: Life's Regulator

The hour is an invention of Western civilization, and it dictates an entire way of life. The hour tells man when to start work and when to quit, when to sleep and when to rise. Man once ate whenever he was hungry, but today mealtimes are fixed to an hour—although not always as strictly as in the picture at right.

Today the hours are such regulators of life that it is hard to think of a day without them. Yet the hour gained importance only in the 14th Century when European towns mounted mechanical clocks that chimed out the time 24 times a day. Prior to that, the day had been divided less accurately. The Anglo-Saxons, for example, broke it into vague sections called tides, such as morningtide, noontide and eventide. But these provided only rough estimates of the time.

As exact as hours can make time, not every society uses them in the same way. In the United States, for example, the office day generally occupies the hours between about 9 a.m. and 5 p.m. In Spain, office workers report at about 9 but take a four-hour break at 1 for dinner and siesta, then return to work at 5 and leave at 8. Spaniards sit down for a light supper around 11 p.m., about the hour most Americans are falling asleep.

EATING ON THE HOUR
Beginning lunch at the same minute every day, these midshipmen at the U.S. Naval Academy in Annapolis, Maryland, are subject to a schedule that governs mealtimes as well as classes and military formations. The practice of holding up empty dishes for a refill streamlines the meal, which can last no more than 20 minutes.

The Demanding Minute

The first clocks had no minute hands; in fact, the minute gained importance only with the development of modern societies. During the Industrial Revolution, trains began to run on schedules, factory whistles blew to change shifts, the tempo of life quickened and the minute became a major governing unit of time.

Even the minute is too long for modern science. In the ventures of man into space, with all the intricate calculations involved, the second, millisecond and microsecond now stand as the vital measures of time.

THE TICKS OF MODERN LIFE
In a railroad station, minutes rule time. Above, travelers watch a clock, waiting for the moment their train leaves. At right, commuters rush for a train that departs at precisely 5:43.

2
Life's Remarkable Rhythms

A biological clock controls the blooming of these flowers of blue-eyed grass, a type of wild iris. Since the stems are strong enough to support only one blossom at a time, one flower blooms each morning and dies that night to make room for the next.

FOR BILLIONS OF YEARS before the first apelike men evolved dim notions of the past, present and future, living organisms were involved with time. Life, from one standpoint, is a perilous sort of equilibrium—a high-wire acrobat perpetually teetering over the gulf of extinction. And to an organism, no less than to an acrobat, timing is crucial.

In a sense, every living thing is a time mechanism. Inevitably, it must run down with time, age and die. To survive, even for a while, it must keep time with itself and its surroundings. Its cells and organs must do the right thing not only in the right place, but also at the right time. A plant that grows leaves before it develops roots will quickly wither; an animal whose heartbeat fails to keep pace with its activity will perish.

An organism must not only synchronize its internal rhythms; it must coordinate them with the outer world. Most plants and animals live in environments that change in regular temporal cycles. A lizard in Mexico's Sonora desert must survive both the blistering heat of noon and the chill of midnight; a bamboo in the Indian jungle must endure both the drumming rains of the monsoon and the baking dry season that follows. Heat and humidity, food resources and enemies, all change as day follows night and summer follows winter. To avoid extinction, plants and animals must at the very least adapt to these external time cycles. In fact, most organisms above bacteria have evolved "biological clocks," which allow them to anticipate the cycles. These mechanisms determine when plants flower and when birds fly south. They also govern the sleeping and waking cycles of men.

The role that timekeeping plays can be seen at every level of life. Inner synchronization appears in one of the most fundamental processes, the division of one cell into two, which requires exquisite timing of dozens of biochemical subprocesses. It also appears, at a far more complex level, in the beating heart of man and other animals.

Everyone knows that the heart's continuous rhythmic pumping is absolutely essential to life. A heart stoppage brings unconsciousness in seconds, death in minutes. To maintain the vital flow of blood day and night, year after year, the thousands of fibers and billions of cells that make up the heart muscle must all work together—and with perfect timing. They do so by virtue of an electrical control system that is unique among the body's organs.

The heart is a two-stage, four-chambered pump. The two thin-walled atria collect the body's blood before it enters the two ventricles. These heavily muscled chambers then drive the blood through the lungs and the rest of the body.

The electrical impulse that governs the contractions of these four chambers begins in the so-called pacemaker, a small lump of tissue in the wall of the right atrium. From there the stimulus spreads over the surface of the atria, causing them to contract. As the blood moves into the ventricles, the electrical impulse reaches a second control center, the atrioventricular node. The node delays the impulse momentarily, giving the atria time to squeeze the last bit of blood into the ventricles. Then the impulse passes on to a special network of fibers in the ventricular muscles and triggers contraction. These fibers are so arranged that the muscles do not contract all at once but in sequence, so that the blood is literally wrung out of the ventricles and into the arteries.

But what keeps the pacemaker ticking on time? Apparently it controls itself, as has been demonstrated by a remarkable series of experiments performed by Isaac Harary and his associates at the University of California, Los Angeles. A few years ago, Harary developed a method of separating living rat heart cells from one another and keeping them alive in a nutrient medium. When he examined the cell cultures under the microscope, he was startled to find that a few of the cells—perhaps one in 100—were beating rhythmically with no outside stimulus.

As these cells grew and multiplied, they sent out filaments to other cells, which formed a network. Then, as these other cells became incorporated into the network, they began to beat in time with the few "leader" cells. Eventually, the thickening network formed itself into fibers, all pulsating together as they do in the living heart. The leader cells, Harary believes, derive from the pacemaker, the "follower" cells from other heart muscle. Thus the heart's pacemaker seems to have a life of its own from the very beginning. Impulses from the nervous system can accelerate it or slow it, but left to itself it will continue to beat.

A heart that beats all by itself

A laboratory experiment familiar to many students indicates what a marvelous timekeeper the heart is. In this experiment a frog's heart that has stopped beating is bathed in a salt solution. The heart starts beating and resumes its natural rhythm by itself.

When we turn from internal timekeeping, whether it is synchronization of cell or muscle, to external synchronization, the biological clocks that keep organisms in time with their environment, we are venturing into dangerous country. Experts argue, not merely about the meaning of experiments—this is common among all scientists—but also about the experiments themselves.

In exploring this tricky terrain, plants are a good place to start. Most of them depend directly on sunlight for life, and should be expected to demonstrate some changes as day gives way to night. And they often do, as men noticed long ago. During Alexander the Great's famous march to India, one of his generals, Androsthenes, recorded that the Indian tamarind tree folds its leaves at night and opens them during the day.

The meaning of Androsthenes' observations did not become clear until the 18th Century, when an inquisitive French astronomer, Jean-Jacques De Mairan, noted that certain species of Mimosa, the so-called "sensitive plants," also open and close their leaves in a daily rhythm. He wondered what would happen if the plants, instead of being exposed to the daily cycle of light and darkness, were kept in continuous darkness. So he shut some mimosas up in a dark shed.

In 1729 De Mairan reported his findings to France's Royal Academy of Science. "It is not at all necessary," he wrote, "that it be in the sun or open air. The phenomenon is merely a little less marked when the plant is kept shut up in a dark place; it still opens out very noticeably during the day and folds up or closes regularly in the evening, for the entire night. The sensitive plant thus senses the sun without seeing it in any way."

Though De Mairan's observation was accurate, his conclusion was wrong: Even the most sensitive plant is not sensitive enough to be aware of the sun without seeing it. What he had unwittingly stumbled on was one of the internal clocks by which plants—and animals—can keep pace with the rhythms of their normal environments even when temporarily isolated from those rhythms.

Flowers that open on the hour

Cyclical daily changes of the kind that fascinated De Mairan are widespread and varied. Flowers often open at a particular hour: The great Swedish naturalist Carolus Linnaeus once planted a timekeeping flower bed which he called "Flora's Clocks." The blossoms of the convolvulus opened at about 3 a.m., the white water lily at 7:00, the marigold at 9:00, and so on for each hour of the day. In many plants, daily rhythms involve not only such particular functions as flowering or leaf-movement but also the most fundamental metabolic processes: the subtle chemistry of photosynthesis, by which they utilize light to manufacture food.

Biologists are now seeking other biological clocks in many plants, using methods that differ little in principle from De Mairan's pioneer experiment. The activity under study—leaf-movement, flowering, metabolism or what not—is first observed under normal day-and-night conditions. If it shows rhythmic variations the plant is then placed under conditions of unvarying light and temperature to see if the rhythm persists. In almost all cases it does, usually for days, sometimes for weeks.

The difficulty, and much of the controversy, surrounding such experiments comes with the attempt to define "unvarying" conditions. The cycles of the plants are extraordinarily sensitive. They can be thrown off by a shift in temperature of less than 2° F. (1.1° C.). A plant kept in constant darkness can have its cycle altered by the few seconds of light needed to observe it.

Even the color of the light can make a critical difference. Some 50 years ago, a group of German experimenters was studying plants under conditions of constant darkness. Once a day, at a regular time, the scientists briefly turned on a photographer's red "safelight" so that they could see to water the plants. The experiment seemed to be giving good results—in fact, results too good to be true, for the plants kept religiously to a precise, 24-hour rhythm. Fortunately, biologists have learned to distrust experiments which give *exactly* the predicted answers. Subsequent studies showed that the red light, though safe for photographic film, was anything but safe for plant experiments, since it acted as a very effective timer that kept the rhythm to an accurate 24-hour period. As the delicacy required

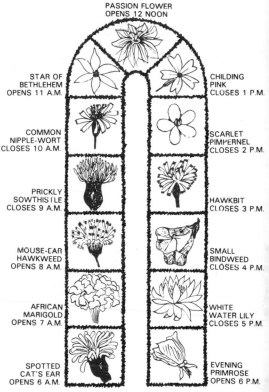

PASSION FLOWER
OPENS 12 NOON

STAR OF BETHLEHEM
OPENS 11 A.M.

CHILDING PINK
CLOSES 1 P.M.

COMMON NIPPLE-WORT
CLOSES 10 A.M.

SCARLET PIMPERNEL
CLOSES 2 P.M.

PRICKLY SOWTHISTLE
CLOSES 9 A.M.

HAWKBIT
CLOSES 3 P.M.

MOUSE-EAR HAWKWEED
OPENS 8 A.M.

SMALL BINDWEED
CLOSES 4 P.M.

AFRICAN MARIGOLD
OPENS 7 A.M.

WHITE WATER LILY
CLOSES 5 P.M.

SPOTTED CAT'S EAR
OPENS 6 A.M.

EVENING PRIMROSE
OPENS 6 P.M.

THE FLOWER CLOCK provided an ingenious and decorative means of telling time in the formal gardens of 19th Century Europe. A series of flower beds was laid out to form the clock "face," with each bed representing a daytime hour. The beds were then filled with flowers known either to open or to close at the prescribed hours. On a sunny day, the time could be determined to within a half hour by this method. Flower clocks are rare today because of the difficulty in finding and cultivating flowers that will "keep time" in various seasons and localities. The examples shown here are selected from flowers common in England and the United States: The times will vary somewhat according to the location of the garden, but the actions of the flowers will always occur at intervals of about an hour.

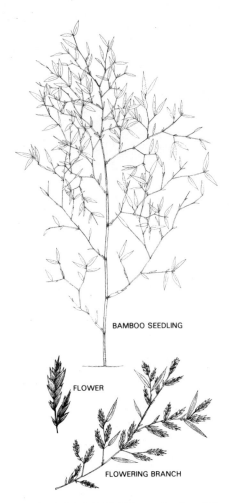

BAMBOO SEEDLING

FLOWER

FLOWERING BRANCH

THE CYCLE OF BAMBOO, the *Guadua trinii* of Argentina, suggests that it has a biological clock as accurate as any yet discovered. This tree's rhythmic life pattern—it takes exactly 30 years from seed to seed—was first studied by agronomist Lorenzo Parodi of Buenos Aires. He transplanted 10 wild seedlings, like that shown at top, which had sprouted in January 1923. The plants grew for 29 years, then produced flowers *(bottom)*. In January 1953 the flowers produced seeds—30 years to the month after the plants grew from seeds themselves.

for such experiments has been recognized, scientists have begun to understand how biological clocks work. There seem to be two basic elements: The first is the clock, which regulates processes within the organism. The second is the "synchronizer," or timer, which like a military cadence caller periodically adjusts the internal clock slightly, prodding to keep it in unison with the external rhythms it must always match. The synchronizer is invariably an external influence—usually light, but sometimes temperature. The clock itself, in the opinion of nearly all investigators, lies within the plant.

Persuasive evidence for biological clocks is the fact that plants under constant conditions rarely operate on a precise, 24-hour cycle. For this reason, the cycles are commonly called "circadian," from the Latin *circa*, "about," and *dies*, "a day." Moreover, plants show consistent individual differences in their cycles: Two bean seedlings raised in the same pot may raise and lower their leaves in cycles of 23 and 25 hours respectively, which is not at all what one would expect if their clocks were operated by an outside mechanism.

How morning glories could bloom at dusk

Such imprecise clocks would seem to be more trouble than they are worth. A morning glory running on a 23-hour clock would gain so much time that in less than a fortnight it would flower at dusk instead of dawn. Fortunately the synchronizer—the normal succession of daylight and darkness—keeps the flower in step with the sun, resetting the clock before its errors reach serious proportions.

The synchronizing effect of light was first studied more than two centuries ago by English botanist John Hill, who reversed the normal cycle by exposing a variety of peas to artificial lighting at night and keeping them in darkness during the day. Hill found that their rhythmic movements quickly adjusted to the new timetable: The revised lighting cycle had reset the clock.

The process of synchronization is still poorly understood. Its importance, however, can be seen dramatically in chicory, the common weed with bright blue flowers.

Outdoors in the meadow, chicory plants normally begin to bloom just before dawn. When placed under constant light, however, they react like an army slowly disintegrating under an overwhelming assault. At first, the divisions lose contact: Each plant drifts into its own circadian cycle, though all its own flowers remains synchronized. Then the regiments begin to break apart: Different clusters on the same plant begin opening at different times. Now disorganization moves down to the company level, with different flowers in the same cluster operating by different schedules. Ultimately the squads lose contact: One or two petals on a flower will open, with the rest remaining tightly closed.

From this and other experiments it seems clear that the clock itself does not reside in any particular organ, controlling the entire plant. Instead, biologists have concluded that each plant cell contains its own subminiaturized clockwork, which independently beats out the time. Only under the influence of an external synchronizer, however, do these millions of tiny clocks begin to "tick" together.

Timing is every bit as important to animals as to plants. Since most animals are more complex creatures and lead more eventful lives, their clocks often serve a greater diversity of functions. Among the most remarkable of these is direction-finding. The sun guides many species, which must possess clocks to tell them when a low sun means east and when west. Bees, for example, navigate to flower beds more than a mile away. The forager bee goes out and flits around until he finds the flowers. Then he flies back to the hive and does an elaborate dance, indicating to the other bees the direction of the food source by its angle from the sun. But these bees can find the food even after the sun has moved across the sky. They do so by means of internal timers that compensate for the sun's movement. They can find their way just as does a boy scout lost in the woods—by comparing the time and the position of the sun.

As might be expected, birds are the most impressive animal navigators. Their direction-finding is remarkably accurate: The migrating Pacific golden plover crosses more than 2,000 miles (3,219 km) of open ocean from the Aleutians—and makes its landfall unerringly in Hawaii.

The way birds navigate

How these migrants perform such feats was until quite recently one of nature's best-kept secrets. Some answers came in the 1950s in a series of ground-breaking experiments. In one, the German zoologist Gustav Kramer, kept starlings in a special circular aviary, lighted by the sun through vertical slits in the walls. During the migrating season, most of the birds perched facing along the compass course that they would have followed had they been free. Kramer set up mirrors outside the illuminating slits which suddenly changed the apparent direction of the sun by 90°—and found that the birds "changed course" by the same amount, clearly indicating that they used the sun as their guide. But when the position of the sun shifted gradually over a period of many hours, the birds, by means of some internal clock, made allowances for the change—as they would have had to do if they were to maintain a straight course on a long flight.

An even more remarkable series of experiments was later performed by another German, E.G.F. Sauer, who proved that some birds can navigate, not only by the position of the sun, but also by the far more subtle patterns of the stars. Sauer found that on a clear night during the migrating season, lesser whitethroat warblers faced southeast, their usual course across the Balkans and Mediterranean to Egypt. Moreover, the birds maintained their course hour after hour, as the stars marched across the sky; like Kramer's starlings, they could compensate for celestial motions by an internal clock. Only when the sky was overcast, with no stars showing, did the birds become disoriented and flutter about at random. Sauer even obtained similar results with a bird that "had spent all its life in a cage and never traveled under a natural sky."

These and other experiments, says Sauer, "leave no doubt that the warblers have a remarkable hereditary mechanism for orienting themselves by the stars—a detailed image of the starry configuration of the sky coupled with a precise time sense which relates the heavenly canopy to the geography of the earth at every time and season. . . ."

The mammal's master clock

The way in which the warbler's chronometer and other animal clocks work remains obscure; so does the working of corresponding mechanisms in plants. But the means by which light synchronizes these clocks in animals is better understood. It does not act directly on the cells as it does with plants, but seems to involve both the nervous system and certain hormone-secreting glands. In mammals and in birds the pineal gland is suspected of acting as the synchronizer and perhaps as master clock as well. This small bit of tissue, buried deep in the brain, was long regarded as one of the body's mystery glands. During the past few years, however, physiologists have determined that the gland's secretions regulate certain periodic activities (such as the sexual cycle in female rats). They are certain that the pattern of secretions shows circadian cycles and that these cycles are regulated by light.

Night and day rhythms are not the only time cycles followed by animals. Many marine creatures dwelling near the shore in tidal waters are attuned to the rise and fall of the tides. Sea anemones open and close, beach worms emerge from or retreat into the sand according to whether the waters are advancing or retreating.

Some species, such as the New England fiddler crab, show both circadian and tidal cycles. Biologists found out long ago that the crab's shell grows dark during the day and pale at night. Some years ago Frank Brown Jr. of Northwestern University, while conducting research at the Marine Biological Laboratory at Woods Hole, Massachusetts, noticed that the maximum darkening seemed to occur a little later every day. By careful observation, he determined that the daily lag was about 50 minutes— almost precisely the same as the lag in the time of low or high tide from one day to the next. The crab's level of activity also seems to be governed by these dual cycles, though in this case the tidal factor appears to be more important than light.

Some marine creatures show physiological changes that are geared to the phases of the moon. These monthly or bimonthly cycles have to do with spawning, and help ensure that both sexes will be sexually active and fertile at the same time.

In many cases, the lunar cycle appears to be governed directly by the monthly variations in the moon's light. In others, it seems to result from a combination of circadian and tidal cycles. Because these cycles are not exactly the same length, they mesh with each other only twice during the lunar month. In some marine animals, an annual cycle of sexual maturity also comes into play, and the interaction of daily, tidal and annual cycles may pinpoint the spawning time so precisely that nearly all the individuals in a given locality will spawn during only a few hours out of the entire year.

Seasonal cycles, which govern such diverse activities as flowering and fruiting in plants, mating and migrating in animals, have a more pronounced effect than the tides or the moon. As one would expect, these rhythms are most widespread in those parts of the earth where the seasonal changes are greatest: the land areas of the temperate and polar regions. There the annual shift from the long, warm days of summer to the short, chill days of winter poses a major challenge to almost every living organism. The need to anticipate these shifts is even more urgent than the need to anticipate the daily shift from day to night. The meadow grass, for example, must begin producing its seeds months *before* the autumn frosts curtail its aboveground activities. The Canadian snowshoe rabbit must start growing its white winter camouflage weeks *before* the first snowfall, or quickly fall victim to an owl or fox.

The long nights that change the rabbit's coat

Despite such stern evolutionary pressures, few organisms seem to have developed anything resembling a "biological calendar," which can count off days as the biological clock counts off hours. Instead, they anticipate the shifting seasons by keeping daily track of the ratio of hours of daylight and darkness, which of course varies as the earth makes its way around the sun.

This phenomenon, called photoperiodism (the response to different periods of light or dark), has been found in both animals and plants. The snowshoe rabbit normally begins its seasonal costume change in August, when the day shortens to less than nine hours. And if, during the long days of July, it is blindfolded for part of the day, the curtailed photoperiod will bring on the changes several weeks earlier.

The physiological mechanisms that enable animals to respond to changing day-lengths are still unclear, though here, as with circadian rhythms, suspicion has fallen on the pineal gland. When it comes to photoperiodism in plants, scientists still do not know what mechanism permits plants to exhibit rhythmic behavior in response to the light-dark cycle. However, the commercial importance of flowering and fruiting is stimulating heavy research in photoperiodic control.

A number of studies have found that the length of night is more crucial than the day length for some plants. The chrysanthemum needs an uninterrupted period of about 13 hours of the darkness of an autumn night before it will flower. Commercial flower growers, seeking to delay the blossoming of the chrysanthemums for a more profitable market, illuminate the plants briefly at night to break up the dark period.

The differences in day or night duration help to explain why some plants flourish in certain areas but not in others. Ragweed, for instance, cannot begin its flowering process until the summer nights have lengthened to about nine and a half hours. In the latitude of Washington,

SOYBEANS' INTERNAL CLOCKS control flowering (and therefore maturation) by measuring the length of daylight; as a result farmers must plant different varieties in different regions. In the Northern states on this map plants must mature by late September or they will be killed by frost. Therefore Northern farmers plant a variety like Chippewa or Hawkeye *(light blue bands),* whose clocks are set to make them flower in the 15-and-a-half-hour days of early July. They mature about two months later, before there is any danger of frost. Farther south the growing season lasts longer, so farmers there choose Clark or Lee varieties *(dark blue bands),* which begin to flower in the 14-hour days of late July and are ready for harvest in late October.

D.C., this occurs around July 1; by the middle of August, the flowers have matured and scattered their pollen, to the intense discomfort of hayfever sufferers. Vacationers in northern Maine or similar latitudes, however, are not troubled, because the ragweed has never established itself so far north. The reason is not that the climate there is too cold for the ragweed to grow—a transplanted plant will flourish. But because the midsummer days are longer in Maine than in Washington, the critical nine-and-a-half-hour night does not occur until August 1, and the plant starts its flowering process too late. Before it can produce mature seeds, it will be killed by the September frost.

In man, as compared with plants and other animals, the biological clock is rather poorly developed. The human animal is as weak in instincts as it is powerful in intelligence—and the instinctive time sense is no exception. To navigate as accurately as the whitethroat warbler, man needs a compass, a sextant and a chronometer. To awaken as promptly as a robin, he usually requires an alarm clock.

Nonetheless, man does show certain circadian cycles that parallel those of the lower animals. Chief among these is the rhythm of sleep and waking. Indeed, barometers of circadian cycles that have been studied— daily variations in hormonal secretions, urine production and body temperature—all seem, under normal conditions, to keep step with the sleep-waking cycle and are probably related to or governed by it.

Resetting man's internal rhythms

Man's sleep-waking cycle seldom runs exactly 24 hours: An English cave expert who spent 105 days alone underground found that he tended to fall asleep a little later every night; his internal "day" averaged 24.7 hours. In some cases, the cycle can be experimentally reset to a slightly different period. Nathaniel Kleitman, an American expert who has studied sleep from many standpoints, proved this during a stay of several weeks in Mammoth Cave, Kentucky, with an associate, Bruce Richardson. The two scientists put themselves on a 28-hour day for 32 days. Their body temperatures, which seem to be the most useful index of the body's circadian changes, were recorded at frequent intervals.

Richardson adjusted to the abnormal days with no difficulty; his temperature regularly hit a peak during waking hours and a trough while he slept. Kleitman's body, however, clung stubbornly to its 24-hour-day cycle; as a result, he was frequently sleepy and irritable during the day and restless at night.

Man's sleep-waking cycle, whatever its individual idiosyncrasies, does not appear immediately at birth. As every parent knows too well, a newborn baby has no fixed pattern of sleep; it will awaken, and demand food, at times that clash with its parents' own sleep patterns. The baby seems to learn its cycle from its parents slowly, over the course of many weeks—in contrast with most cycles in the plant and animal worlds, which require only a day or two to establish themselves.

Man's internal clock remains flexible through life, adapting easily to gradual change. As early as 1907, an English doctor on a voyage from Australia to England noted that his cycle of temperature variation kept pace with the ship's passage through different time zones. It regularly reached its daily maximum at 6 p.m. local time, though by the time he reached London, this corresponded to 4 a.m. Melbourne time.

The jet set's mixed-up clock

The human clock cannot, however, reset itself to today's jet speeds. In the six-and-a-half-hour flight from New York to London, a traveler will get himself five hours out of step with the daily rhythm of activities and will probably feel tense, tired and irritable. Studies carried out by the Federal Aviation Agency show that these feelings are not imaginary; a rapid shift across time zones produces consistent and measurable disruptions of physiological and psychological processes. A group of volunteers was flown from Oklahoma City to Manila (a time difference of 10 hours) and then tested. During the first 24 hours after arrival none of them could concentrate long enough to add a column of 10 two-digit numbers. For some, their "reaction time" (which is of crucial importance in driving a car safely) more than tripled. Similar experiments with a north-south flight of equal length, in which the time difference was an hour or less, produced no functional impairments.

The "time-zone syndrome" is even more important in the case of airline pilots, who seldom have time to adjust completely before winging back to their original time zone—and whose job requires the utmost in mental alertness. Many of them definitely prefer north-south to east-west flights; says one: "Flying from New York to Buenos Aires, my digestion can still work on Eastern Standard Time!"

It takes at least 24 hours for mental alertness to return to normal after a long flight across time zones. Indeed, the body's temperature cycle does not adjust to the "new" time for several days longer. As a result, FAA psychologists have recommended that tourists flying east or west should rest for a day before embarking on strenuous sightseeing programs, and that executives or diplomats should not undertake any important business in the first 24 hours after arrival.

Scientists studying human circadian rhythms are experimenting with drugs that can be used to reset biological clocks, with the hope that they may one day be able to prescribe exact drug dosages to help ease the special problems of jet travelers and workers changing shifts.

Unlike plants or other animals, man appears to lack built-in seasonal rhythms. Perhaps, having evolved in the tropics, "where winter never comes," he never developed them. He does not migrate with the birds or hibernate with the rodents. His metabolism, to be sure, changes somewhat with the seasons, but it does not appear to anticipate them. A warm, moist April day, impinging on our still-"winterized" bodies, brings on the delicious languors of spring fever year after year.

Nor, for that matter, does man mate or spawn at any special time of year. In these as in his other activities, he remains a man for all seasons.

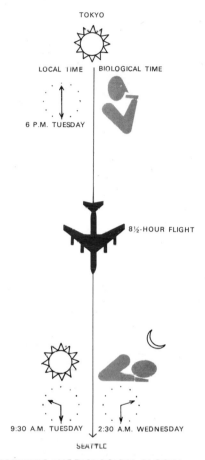

RESETTING THE BIOLOGICAL CLOCK is a problem for travelers, who may fail to adapt to local time for several days after a long trip. On a flight from Tokyo to Seattle, the traveler who leaves Japan at 6 p.m. Tuesday may already have eaten dinner *(top right)*. His biological clock is perfectly synchronized with local time *(top left)*. He then flies east for 8 and a half hours, crossing six time zones (as well as the International Date Line), and arrives in Seattle at 9:30 Tuesday morning local time *(lower left)*. But this biological clock *(lower right)* is still tuned to Tokyo time and he feels as if it were 2:30 a.m. Wednesday. Instead of being wide awake and ready to face the day, he is badly in need of a good eight-hour sleep.

The Biological Clock

To anyone who flies to Europe or across the United States, the time dislocation of jet travel is a familiar annoyance. The day seems too short or too long; it is midnight when it should be dinnertime, or 9 o'clock in the morning when it feels like 2 a.m. The traveler's realization that he is suddenly out of step comes from his internal timekeepers, which, in man as in other living things, regulate activity. Each organism has certain natural rhythms that determine how it lives and grows and, in some cases, whether it survives.

Many of these rhythms are self-contained, but others must be synchronized with the outer world. The biological clocks that accomplish this, like kitchen clocks, have two main characteristics: They are powered internally, and they are adjusted to their own time cycles by some external cycle — usually changes in temperature, the rising and setting of the sun or the tides. Beyond these facts, little is known about nature's mysterious timekeepers. In laboratories around the world, scientists are exposing plants to varying cycles of darkness and light, analyzing the chemistry of animals and subjecting men to strange sleeping-waking patterns — trying to learn more about biological clocks and what makes them tick.

A MICROSCOPIC TIMEKEEPER
The one-celled red-tide alga reproduces on a regular schedule by dividing itself into two new cells as shown in this scanning electron microscope photograph, magnified 3630 times. Each cell divides at least once in every three-day period, invariably at a "peak" time that occurs at the end of a 12-hour dark cycle and the start of a 12-hour light cycle.

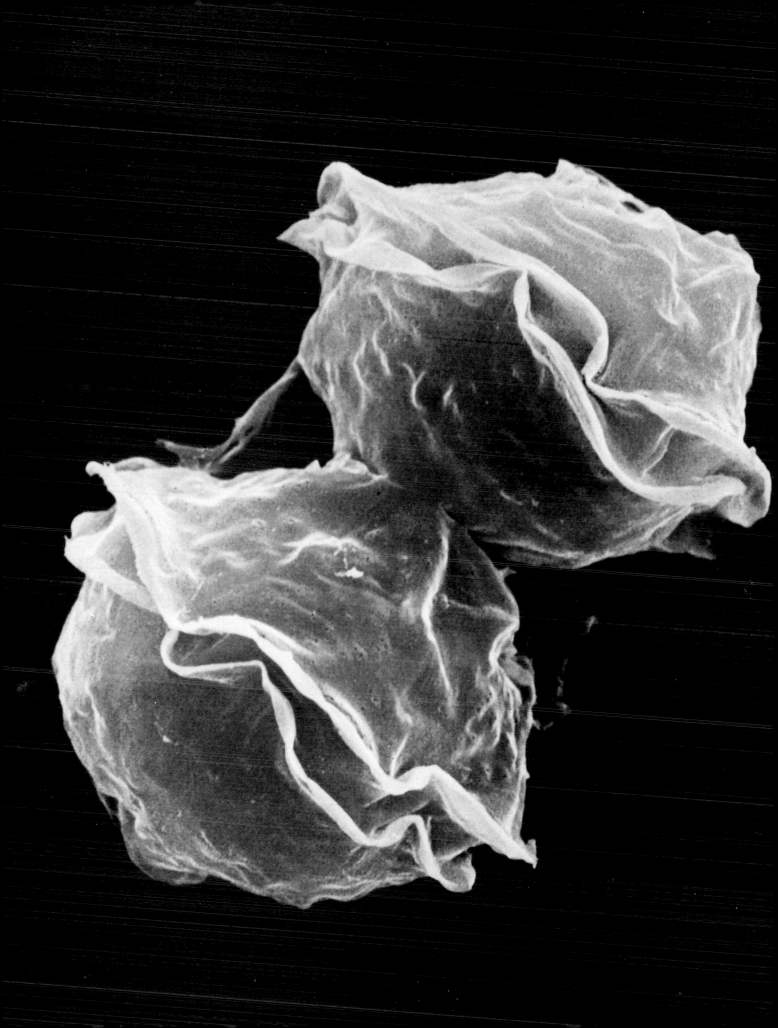

A 24-HOUR CYCLE

Cut blossoms of the kalanchoe plant keep their regular day-and-night rhythm despite the absence of daylight in a laboratory at the Botanical Institute in Tübingen, Germany. In the top picture the blossoms are open while it is daytime outside the lab. Below they have closed while it is night outdoors. The blossoms' openings and closings are noted automatically: The shadows cast by their petals move over photoelectric cells set in the circular depressions. The cells' varying output is converted by a recording device into a graph of petal movements.

Flowers That Bloom on Schedule

Many flowers open and close on a schedule almost exactly like that of the sun's rising and setting. In order to approximate this cycle so closely, they must contain timing devices of some sort. The thick-leafed kalanchoe plant *(left)* and the Biloxi soybean *(right)* are striking examples.

The kalanchoe ordinarily opens its blossoms in the morning and closes them every night—and it continues to do so even when kept in the dark. To maintain this regular rhythm without any external signal such as a change in illumination, the plant itself must be able to count the hours of the day. Where this internal clock mechanism might be located and how it might operate are under investigation by Erwin Bünning of the Botanical Institute in Tübingen, Germany, who records the opening and closing of the blossoms with special laboratory devices. The results, Bünning says, rule out a specialized timing organ and point to the living material inside each cell as the clock.

A far more complex timekeeper is under investigation at the University of California at Los Angeles by Karl Hammer, who has been working with the Biloxi soybean. This plant behaves almost like a calculating machine: It will bloom only when exposed to periods of light and dark in cycles that are multiples of 24 hours. The plants require at least eight hours' light, but the period of darkness can be extended provided that light and dark hours add up to a total of 48, 72 or some other multiple of 24. Even a slight disruption of the cycle —a brief flash of light during the required dark period, for example— prevents the plants from blooming.

CLOCKING A LEAF MOVEMENT
In a laboratory at U.C.L.A., Karl Hammer studies an internal clock of a Biloxi soybean plant by timing movements of its leaves, which rise and fall on a 24-hour schedule. A wire connects a leaf to a metal stylus, which records the frequency of leaf motion on a revolving cylinder.

Creatures That Live by the Clock

Without biological clocks, many animals would not survive. The monarch butterfly dies if it emerges from its chrysalis too early in its development. The flealike sandhopper evidently depends on two distinct timers for its survival. One tells it when to come out of its burrow, the other helps guide it toward food.

The clues to these two clocks come from the tiny creature's regular habits. Heat-sensitive sandhoppers, even in the laboratory, usually leave their burrows after the sun has set. Occasionally, however, they do venture out in daylight—and when they do they always search for food in one direction. Their daytime foraging seems to be guided by the position of the sun. Since the sun's position changes, they must determine direction with the aid of an internal timer.

A MONARCH'S METAMORPHOSIS
An inner time sense controls the development of the caterpillar at upper left into the monarch butterfly at bottom right. This three-week-old caterpillar *(picture 1)* builds a chrysalis *(pictures 2-4)* where it rests 10-12 days before emerging *(pictures 5 and 6)* as a mature butterfly.

STUDYING THE SANDHOPPER
To observe the nightly movements of sandhoppers, researcher Floriano Papi *(left)* of the University of Pisa and an assistant set up a timing device on a beach near Pisa. Most of the sandhoppers come out after dusk, but sometimes they appear during the day. When put down on dry sand away from the sea, they always head west, toward the water, navigating by means of the sun and their biological clocks.

SEASIDE CLOCK WATCHERS
Sandhoppers cluster in a scientists' timing device, shown here with its cover removed. As they move along the sand, they fall through a hole in the cover into one of the 24 pie-shaped compartments in the plastic dish. A clock mechanism turns the dish so that one compartment passes under the hole each hour. The number of sandhoppers in each compartment shows when these creatures are most active.

A vigilant researcher, armed with a telescope and other sighting devices, scans Tortuguero Beach in Costa Rica for nesting green turtles.

The Mystery of Turtle Navigation

Not all biological clocks are as easily understood as those of men and butterflies. One of the most mysterious belongs to the green sea turtle. To reach their accustomed mating sites, which are hundreds of miles from their feeding grounds, these turtles find their way across stretches of empty ocean that would challenge a human navigator armed with charts, compass and chronometer. Scientists reason that this feat must require complex internal clocks so that the turtles can time the sun's movements in the sky and determine direction from its shifting position. But the working of these clocks is unknown.

The precision with which green turtles periodically cross the ocean is unerring. One Brazilian colony swims to Ascension Island in the Atlantic to nest, traveling 1,300 miles (2,092 km) to a target five miles (8 km) wide. Every July, Tortuguero Beach in Costa Rica hosts turtles that have migrated 800 to 1,000 miles (1,287 to 1,609 km) from Nicaragua, Panama, Mexico and Florida. Scientist Archie Carr of the University of Florida, who spent many summers tagging and tracking nesting females on Tortuguero Beach, discovered that each female returns to Tortuguero to nest every two or three years and that usually she will crawl out of the ocean to dig a nest which is within 200 yards (183 m) of all her previous nests.

Carr believed that the turtles' precision navigation may be accomplished by a number of innate direction finders. Although they are notoriously nearsighted above water, the turtles may be able to gauge their position, as birds do, by reading the sun or the stars. Distinctive coastal smells, borne through the ocean on currents, may also guide them to their targets.

DIGGING A SANDY NEST
A green turtle on Australia's Great Barrier Reef sends the sand flying as she makes her nest. The sea often is not visible from the nest sites the turtles choose. Nonetheless, when the young hatch they crawl straight to the water, another sign of an innate biological compass.

MAKING STRAIGHT FOR THE SEA
Her nest completed, a green turtle leaves a tractor-like trail on Tortuguero Beach as she heads into the sea. Male turtles remain in the ocean shallows during the nesting season, which lasts from one to three months. Females come ashore about every 12 days to lay eggs. At season's end all navigate together hundreds of kilometers back to the feeding grounds.

The Human Rhythms

Biological clocks in man are subtle, but a number of undeniably clocklike rhythms are present. The pulse normally beats 75 times a minute, the lungs fill with air at a regular tempo, even the brain thinks more clearly at certain times of day.

Many of the body's rhythms seem to be timed by the 24-hour clocks which predominate in nature. Man needs a regular—usually nightly—period of rest. And even when he sleeps, the electrical impulses from his brain have their own special rhythms (right).

Longer cycles are also evident in man. The female menstrual cycle is probably a vestige of the lunar schedule followed by many marine creatures. What may be a multiple lunar cycle shows up in the 120-day average lifetime of red blood cells.

However they are timed, all the body's rhythms must work in concert. If the cycles get out of step—as they do when a factory worker changes from a day to a night shift—the resulting strain can sometimes lead to irritability and lapses in judgment.

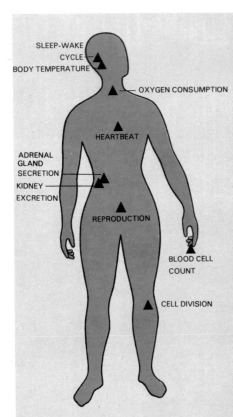

SLEEP-WAKE CYCLE
BODY TEMPERATURE
OXYGEN CONSUMPTION
HEARTBEAT
ADRENAL GLAND SECRETION
KIDNEY EXCRETION
REPRODUCTION
BLOOD CELL COUNT
CELL DIVISION

THE SLEEP-WAKE CYCLE, controlled by the brain, is the most familiar biological rhythm in man; it influences the timing of many of the body's other functions.

BODY TEMPERATURE, regulated by the hypothalamus, varies each day, regularly falling to its lowest point during normal sleep, between 1 a.m. and 7 a.m.

OXYGEN CONSUMPTION increases during the normal peak hours of the body's activity—whether or not activity is present.

HEARTBEATS normally occur 75 times a minute, but the rate drops slightly to a low point between 10 p.m. and 7 a.m.

ADRENAL GLAND SECRETION dips during sleeping hours, but rises before waking to prepare the body for the day's activities.

KIDNEY EXCRETION of such waste products as potassium and sodium reaches its peak during the middle of the day.

REPRODUCTION is dependent on the periodic production of ova by the ovaries, in cycles of about 28 days. Pregnancy usually lasts 270 days from fertilization to birth.

BLOOD COUNT has its own rhythms: The number of red and white corpuscles falls to a daily low in the morning hours.

CELL DIVISION, occurring throughout the body, takes place most frequently during the late evening.

THE BODY'S TIME CYCLES
Biological rhythms ebb and surge throughout the body: Some of the most important ones are located in this diagram. Many of these rhythms are so pronounced that time could almost be measured by them—the minutes by the pulsating heart, the days by the rising and falling of body temperature or of adrenal secretion, the months by menstruation cycle and pregnancy.

RECORDING THE SLEEP CYCLE
Strung with wires like a living marionette, a volunteer at the University of Chicago rests soundly while his eye movements and brain waves are captured electronically on an electroencephalograph. The resulting brain wave patterns are determined by the depth and duration of sleep.

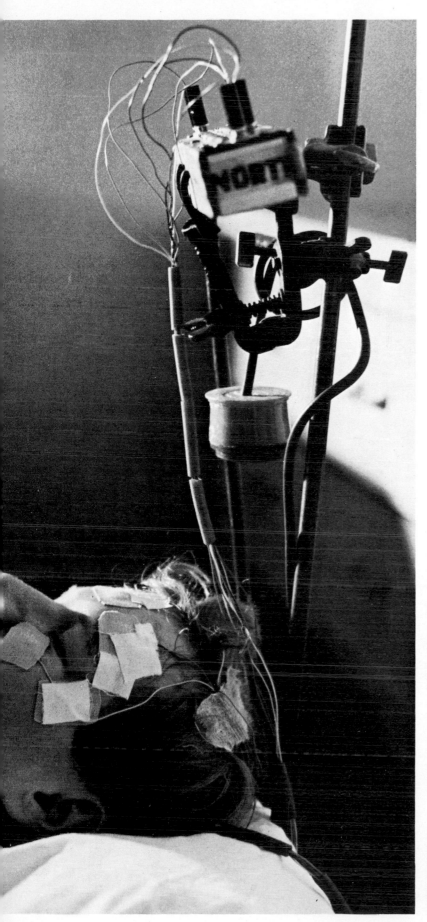

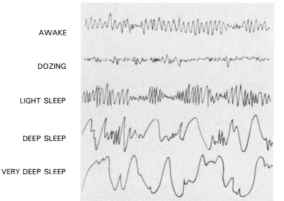

AWAKE

DOZING

LIGHT SLEEP

DEEP SLEEP

VERY DEEP SLEEP

BRAIN WAVES WHILE SLEEPING

An electroencephalograph *(below)* records the changing strength and frequency of brain waves as wiggles in a chart *(above)*. In the top line, the patient is awake. Then, in the first stages of sleep, as muscles relax and thoughts wander, the waves grow faint and irregular *(second line)*. After about 10 minutes, the waves assume a more definite pattern. A half hour later, they trace the hills and valleys of deep sleep, reaching peak intensity *(bottom line)* after 90 minutes.

Resetting a Human Clock

Evidence for biological clocks stems not only from the way plants and animals can keep track of time when removed from the normal day-night cycle, but also from the ability of many living things to reset their timing.

A classic demonstration of this occurred in 1938 when Nathaniel Kleitman and Bruce Richardson of the University of Chicago stayed for 32 days in a 60-by-20-foot (18-by-6-m) chamber of Mammoth Cave, Kentucky. Kleitman and Richardson tried a 28-hour cycle, staying up 19 hours and going to bed for nine. Within a week, Richardson, then 23, had reset his biological clock, but Kleitman, 20 years older, was so firmly tied to the 24-hour cycle that he could not make the adjustment.

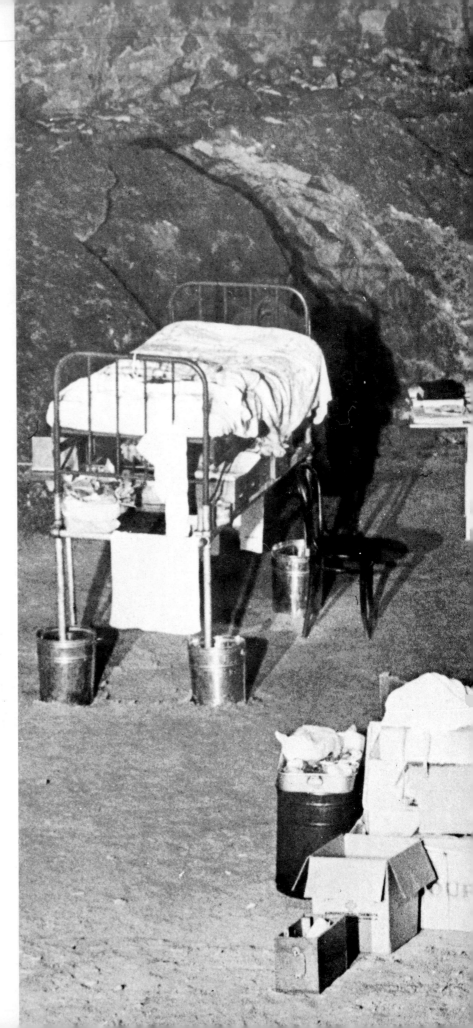

SCIENTIFIC CAVE DWELLERS
Experimenters Nathaniel Kleitman *(left)* and Bruce Richardson emerge from Mammoth Cave after their pioneering test of the human clock's constancy. During their month-long stay in the cave *(right)* they occupied themselves by studying and by keeping records of their reactions to life on a 28-hour cycle. Their observation that some people can adapt the body's timekeepers to new schedules has since been confirmed by other cave-dwelling scientists.

3
Subdividing the Year

Stonehenge, seen at twilight from a point on its main axis, may have timed the beginnings of seasons over 3,500 years ago. The moment of the first sunrise of summer is still pinpointed from Stonehenge by sighting along this axis.

TO THE LOVER OF OLD MOVIES, one of the most familiar screen images is a close-up of a calendar, its leaves peeling off one by one to symbolize the passage of time. Trite as it may seem, this Hollywood cliché makes its point effectively by employing one of civilization's most familiar and basic tools. The calendar that hangs on the kitchen wall is an indispensable part of everyday life. It enables man to keep track of passing time and to plan for the future. It tells him when to plow or plant, when to prepare for fasts or festivals, when to lay in fuel for the winter. It warns him when to plan his son's birthday party and lets him know how many shopping days there are until Christmas.

Of equal importance, the calendar enables a man to coordinate his activities with those of other men thousands of miles away. Since the same calendar hangs on all the walls and sits on all the desks (in Western countries, at least), a man from Chicago can make a business appointment in Los Angeles or a hotel reservation in Paris with the assurance that both he and his far-off correspondents are talking about the same day.

The calendar is a deceptively simple device, a means of counting the days and organizing them into conventional units—years, months and weeks. These units generally derive from recurrent astronomical cycles, which are the most regular and among the most conspicuous changes in nature. The day, of course, is based on the earth's 24-hour rotation on its axis—although time can now be measured independently with such accuracy that it is possible to tell when the earth is a little slow or fast. The month, and probably the week, come from the phases of the moon, while the year stems from the tilted earth's revolutions about the sun. Observing these cycles and measuring them may seem simple enough, but organizing them into practical time-reckoning systems has taxed man's scientific resources for thousands of years. Even today, we do not—and cannot—have a completely accurate calendar; we merely have one that is good enough to serve our needs.

If astronomical observation has played a major role in the development of the calendar, so has religion. Through history, man's tendency to worship heavenly bodies as well as observe them has given calendar making a religious significance. The names of our months enshrine the Roman gods Janus and Mars and the goddesses Maia and Juno. The days of the week commemorate the Teutonic deities Tiw, Woden, Thor and Fria, as well, of course, as the sacred Sun and Moon. For centuries upon centuries religious dogmas and priesthoods have been involved with time-reckoning, reshaping old calendars and stimulating or blocking the adoption of new ones.

Both the scientific and religious threads in the history of the calendar stem from man's need to anticipate the future and to prepare for it. His body's built-in rhythms—unlike those of many lower organisms—do not keep him in time with the seasons. His mind has had to anticipate changes in nature in order for him to survive. Keeping track of time, however, did not begin with counting the days in the month and the

SUNDAY—Sun's day, from Latin *Dies solis*

MONDAY—Moon's day, from Latin *Dies lunae*

TUESDAY—after Tiw, Teutonic god of law

WEDNESDAY—after Woden, principal Teutonic god

months of the year. Long before man learned to ask "How many?" he began to wonder "When?" When will the migrating reindeer or mammoth pass through our valley, so we may have our spears and pitfalls in readiness? When will the winter storms burst upon us, so that we can take timely shelter? In asking such questions, men no doubt gradually learned to associate significant changes in their environments with changes in the heavens.

Some observers believe the moon was man's first means of keeping track of time. One writer, after analyzing a great number of curious prehistoric markings, some painted on pebbles, some scratched on bone, has concluded that they are tallies recording the days of the lunar month, from one new moon to the next. Some of these Paleolithic patterns may be as much as 35,000 years old; if he is right, they would seem to establish the lunar method as man's first time-reckoning system.

Telling time by the stars

Most historians of astronomy, however, do not agree with this view. Studies of primitive societies still existing have provided strong evidence that the stars are more widely used for time-reckoning than the moon. Heliacal risings or settings—the appearance of certain stars near the horizon just before sunrise or after sunset—are a favorite indicator. One primitive tribe in Australia, for example, determines the proper season for hunting termites by noting the position of the bright star Arcturus on the horizon in the evening. Another bright star, Vega, tells the tribesmen when it is time to search for the eggs of the mallee hen.

Primitive agricultural peoples are especially dependent on the stars. The Tukano Indians of northwest Brazil, for example, reckon the seasons by watching the Pleiades. When this cluster of stars dips below the horizon just after sunset, they know it is time to plant their crops, since the heavy seasonal rains are about to begin.

For many primitive tribes, it has been only a small step from the observation that the shifting motions of heavenly bodies indicate the seasons to the notion that these same bodies control the seasons—and must be propitiated if the tribe is to prosper. The Bakongo in central Africa speak of the Pleiades as "the-caretakers-who-guard-the-rain." The primitive Bushmen of southern Africa hurl flaming torches at the "Winter Stars," Canopus and Sirius, and chant incantations to them to make them rise higher in the sky and bring the winter to an end. Indeed, millions of civilized people today follow astrology—a belief that the stars bring good fortune or bad.

Even in primitive societies time-reckoning is a complex task. The Kenyah tribes of Borneo keep track of the seasons by observing changes in the height of the sun, and their methods for doing so are sufficiently complicated so that they must be carried out by a group of specialists. According to the anthropologist Martin P. Nilsson, "The determination of the time for sowing is so important that in every village the task is entrusted to a man whose sole occupation is to observe the signs. He need not cultivate rice himself, for he will receive his supplies from the other inhabitants. . . . The process is a secret, and his advice is always followed."

Once secret methods of time-reckoning were combined with the ancients' belief in the need to placate the heavens, a key figure emerged: the priest-astronomer who observed the heavens, interpreted them to lesser men, and by his incantations and sacrifices purported to ensure that their influence would be benign.

The fusion of religion and time-reckoning among primitive peoples achieved its most impressive expression in the megalithic ("big stone") cult of northwestern Europe, whose circles and avenues of standing stones still dot the landscape in parts of England, Brittany and Scandinavia. Erected by communities of farmers and herdsmen in the centuries following 2000 B.C., these structures undoubtedly were places of worship, concerned also with reckoning the seasons.

Stonehenge: An ancient way of reckoning time

The largest of these monuments is Stonehenge, in southern England. Even today, half-ruined by the passage of some 35 centuries, Stonehenge is an impressive sight, its gigantic roughhewn stone pillars looming against the sky over the broad expanse of Salisbury Plain. When it was complete, Stonehenge included four main structures. The outermost of these was a circular colonnade, 98 feet (30 m) across, of 18-foot (5.5 m) sandstone pillars, or sarsens, roughly squared and capped with equally rough lintels. Within this group of pillars was another ring of smaller, more crudely shaped stones. Inside the two circles were two horseshoe-shaped stone formations.

These inner structures were the temple proper. Time-reckoning required three other elements: the altar stone, centered in the open end of the two horseshoes; a pair of large sarsens, standing to the northeast of the main sarsen circle; and beyond that pair still another sarsen, known as the heel stone.

For the builders of Stonehenge, the chief festival, and perhaps the beginning of the year, was Midsummer's Day (June 24). At dawn on that day, the high priest could stand in the center of the monument at the altar stone, peer through the pillars of the two great circles and the two outer sarsens, and see the rising sun just above the heel stone. At midwinter, near the year's shortest day (December 22), he could gaze out in the opposite direction in the evening and see the setting sun between the two outer sarsens.

Certain other megalithic temples appear, like Stonehenge, to celebrate a year running from one Midsummer's (or Midwinter's) Day to the next. Still others are built around the planting year, beginning early in May. Both types of temples were dedicated to the worship of the sun's life-giving warmth and power, and although it cannot definitely be proved, both types probably served also to keep track of the seasons by solar observations. Even so complicated a structure as Stonehenge

THURSDAY—after Thor, Teutonic god of war

FRIDAY—after Fria, Teutonic goddess of love

SATURDAY—after Saturn, Roman god of agriculture

THE NAMES OF DAYS in English have a mixed ancestry. All were inherited from Teutonic tribes, but the Teutons had borrowed some names from the Romans. Thus, Sunday and Monday, honoring the sun and moon, are Roman; so is Saturday, commemorating the Roman god Saturn. For other days of the week, the Teutons discarded Roman gods for their own gods—so the fourth day of the week commemorates the Teutonic god Woden. (In French and Italian, however, that day honors the Roman god Mercury—thus, the French *mercredi*, the Italian *mercoledi*.)

was not a calendar; at best it served to let its builders know when certain religious and agricultural tasks must be performed. It did not divide the year into units. There is no evidence that its builders knew—or cared—that the solar year they celebrated contained a particular number of days.

Knowledge of this sort almost certainly required observation and record-keeping over a fairly prolonged period. This implies a society in which some people have not only learned to keep records but make it their business to do so.

The first great calendar makers

The earliest clear example of such a society occurred about 5,000 years ago on the banks of the Tigris and Euphrates Rivers, among the Sumerians, that singularly gifted people who evolved the first literate, urban culture. The Sumerians had the specialists for calendar making: priestly scribes who kept records on tablets of damp clay and who, doubtless, had already established themselves as professional time-reckoners.

In each of the small Sumerian city-states, the priests were responsible for administering the land on behalf of the gods and of the gods' earthly representative, the king. The job was not a simple one. To build a civilization out of the mixture of swamp and desert that is lower Mesopotamia required a network of dikes and ditches for drainage and irrigation. The construction of these works, and of the elaborate mud-brick temples over which the priests presided, required the coordinated labor of scores of men. Once built, the drainage and irrigation systems had to be kept in repair. Most importantly from a time-reckoning standpoint, the irrigated fields of wheat and barley, onions and cucumbers, had to be plowed, sown, tended and harvested at definite times of the year. Market days were held periodically in each of the small towns in the kingdom. The gods, on whose goodwill the prosperity of the kingdom depended, had to be appeased with prayer and sacrifice on certain holy days—and these ceremonies had to be held on the same day in each town. For such a highly complex society a rough reckoning of the seasons would not do.

Our knowledge of the Sumerian calendar, unfortunately, is limited; it comes mainly from documents of the Babylonians, who succeeded the Sumerians as lords of Mesopotamia. It seems fairly certain, however, that the Sumerian priests based their original calendar on the moon, dividing the year into 12 lunar months of 30 days each. This arrangement confronted them with astronomical problems that have plagued calendar makers for thousands of years.

The main problem stems from the fact that the astronomical cycles from which the day, the year and the months are derived do not fit neatly together. The year, which is based on the earth's revolution around the sun, comes to about 365¼ days. The month, of course, is measured by the phases of the moon, the full cycle coming to a little more than 29½ days. As a result, the year is not composed of 12 equal months, but of about

"HEY, SAM, THE BIG ROUND YELLOW THING CAME UP AGAIN."

THE WONDER OF THE SUN has fascinated man ever since he dwelt in caves, as this cartoon recalls. For thousands of years he believed that the sun revolved around the earth, and he counted on its regular appearance to mark the passage of his days.

12⅓. Unless the proper corrections were made, the Sumerian calendar, with its 30-day months, would have rapidly got out of step with the moon and the sun. The Sumerian priests must have corrected the calendar, but their exact methods have been lost. We do know, however, that their Babylonian successors were able to keep the months in step with the moon by alternating 30-day months with months of 29 days, and adding an occasional extra 30-day month to make up for lost days. Similarly, they kept the year in step with the sun by throwing in an extra month every three years or so.

For many centuries these calendar corrections were a matter of cut-and-try. A letter of the great Babylonian King Hammurabi, written some 1,700 years before Christ, says that the King ordered the insertion of an extra month whenever he happened to notice (or a priest called it to his attention) that "the year hath a deficiency." To the Babylonians, as well as the Sumerians and their prehistoric predecessors, the heavenly bodies were manifestations of the gods; if their motions seemed a bit capricious, who was to question the gods' actions? The best that men could do was to keep track of the motions of the heavenly bodies and pray that the gods would hold their caprices within bounds.

The race between the sun and moon

The cumbersome Babylonian system of correction was followed for at least 15 centuries. By around 500 B.C., however, the priests had discovered by lengthy observations that a certain order prevailed in the seemingly erratic motions of the sun and the moon. Every 19 years the solar and lunar cycles assumed the same "phase relationship" to one another. This relationship can be understood by likening the sun and the moon to a heavy truck and a racing car traveling around a circular track. The truck is the sun and the car is the moon. While the truck makes one circuit (a year), the car will make 12.37 circuits (lunar months). The car, of course, will pass the truck frequently, and at various points on the track. (The car-passing is equivalent to the appearance of a new moon.) When the truck has completed 19 circuits, however, the car will pass it at almost the same point where both vehicles started. Then the whole cycle will begin again.

It was this 19-year phase relationship between the sun and the moon that the Babylonian astronomers discovered. They found that 19 solar years were almost exactly equal to 235 lunar months. With this information they could work out the first rational system of calendar correction, providing for extra months in seven specified years out of the 19, making a total of 235 months.

The 19-year cycle is now called the Metonic cycle, after Meton, a Greek who may have brought it home from Babylon. The cycle had only a marginal effect in Greece, where a calendar akin to the earlier Babylonian one was in use, each city-state having adopted it in a slightly different form. Being as contentious in calendar making as they were in politics, the Greeks, even with the help of the Metonic cycle, never suc-

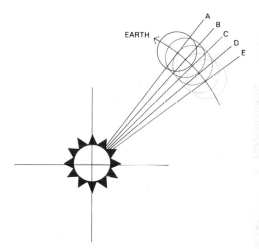

THE NEED FOR LEAP YEAR stems from the fact that the time the earth takes to orbit the sun does not add up to an exact number of days. The diagram above shows what would happen without leap year. If the earth begins an orbit at point A, it should return to A in a year. But 365 days later it is at B, almost a quarter day short of its starting point. Two calendar years later the earth is at C, three years later at D and after four years at E, a full day short of point A. In a century, New Year's Day would slip back more than 20 days. To correct this cumulative error a day is added each leap year.

ceeded in adopting a unified calendar for their whole country. However, with the conquest of Greece by Rome in 87 B.C. to 84 B.C., the matter became academic, since civil affairs from then on were regulated by the Roman calendar. This calendar is the direct ancestor of our own.

The story of the Roman calendar, like that of the lunisolar calendar, goes back to another ancient river-valley civilization, this one along the Nile. The Egyptians, whose civilization seems to have been quickened to life by Sumerian influences, were less advanced technologically than their Mesopotamian contemporaries. Their calendar, however, was better. Though they seem to have used a lunar calendar for certain religious purposes (Thoth, their moon god, was also the divider and reckoner of time), they soon abandoned the task of trying to reconcile solar and lunar cycles and settled on an unvarying 365-day year.

The star that came with the flood

In part, this decision may have resulted from a coincidence. By far the most important seasonal event along the Nile was the river's annual flood, which watered the fields and fertilized them with silt, making Egypt, in Herodotus' famous phrase, "the gift of the Nile." Around 3000 B.C., when the Egyptian priests were shaping their 365-day calendar, this crucial event coincided with a conspicuous astronomical happening. In mid-July, when the flood began, Sirius, the brightest of the fixed stars —called Sothis by the Egyptians—appeared in the east just before sunrise. The star was particularly noticeable because for several weeks previously it had been above the horizon only in the daytime, when it could not be seen. As a result of this coincidence, the Egyptians concluded that "the rising of Sothis" marked the arrival of the flood—and the beginning of the year.

The Egyptian priests, like the Sumerians, divided the year into 12 months of 30 days each. But instead of worrying about extra months, they simply tacked on an extra five days every year. These "days of the year" were spent in feasting and religious services honoring Sothis and celebrating his watering of the land.

The first recorded attempt to improve the Egyptian calendar did not come until 238 B.C., by which time the land of the Pharaohs was ruled by a foreign dynasty, the Greek-speaking Ptolemies. One of these monarchs, Ptolemy Euergetes, proposed that the civil and Sothic years be harmonized by the addition of a sixth "day of the year" (the equivalent of our February 29) every fourth year. But the conservative Egyptian priesthood, with 25 centuries of tradition behind them, ignored the edict —and Ptolemy was wise enough not to press the point. Calendar reform had to wait another two centuries. The man who finally put it across was that remarkable general, historian and politician, Julius Caesar.

When Caesar came to power, the Roman calendar seemed hopelessly confused. The Romans had never cared much for science (though they were fine engineers), and their priests had done a very amateurish job of keeping their 355-day lunar calendar up to date. Politicians had

CALENDAR AUTHOR Aloisius Lilius was a noted professor of medicine at Italy's University of Perugia who spent 10 years working out the details of what later became known as the Gregorian calendar. The new calendar was presented to the Roman Curia in 1576 and was debated by an eminent committee of clergymen and scientists appointed by Pope Gregory XIII. The reform was finally promulgated by the Pope in 1582.

added to the confusion with "corrections" that were designed to extend their own terms in office or to curtail the terms of their opponents. By Caesar's time, the calendar was more than two months out of step with the seasons.

Caesar may have become acquainted with the Egyptian calendar on the same trip during which he got to know Cleopatra. He certainly came to know the Greek-Egyptian astronomer, Sosigenes. On the advice of Sosigenes, Caesar decreed that the year 46 B.C. would be 445 days long, containing an additional 23 days at the end of February and 67 days between the months of November and December. In Roman tradition, that year went down as "the year of confusion," but the year was now back in step with the seasons. To keep it there, Caesar ordered that every fourth year the month of February should have an extra day—the first "leap year."

It was to this Julian calendar that the civil affairs of the Roman Empire—and after Rome's fall those of the civilized parts of Europe—were attuned for some 16 centuries. Religious affairs, however, were a somewhat different story. The Christian Church, which took over most of the Jewish Scriptures, also took over parts of the Jewish calendar. This was the complex lunisolar calendar, which the Jews, in turn, had taken over, Metonic cycle and all, from the Babylonians.

The Church scheduled certain of its feasts, such as Christmas and the saints' days, on fixed dates. Others were "movable feasts," or varying dates, some of which were directly related to the Jewish calendar. The chief among these was Easter, which according to the Scriptures fell at the same time as the Jewish Passover. At the Council of Nicea, in 325 A.D., the Church dignitaries, after considerable argument, established the modern method of dating Easter. They placed it on the first Sunday after the first full moon after March 21 (the spring equinox).

How the seven-day week began

Among the features of the Jewish calendar adopted by the Christians was the seven-day week. Originally, the week may have been merely the interval between market days; many semicivilized peoples even today observe "weeks" of this sort, ranging in length from four to 10 days. The religious significance of the seventh day, as expressed in the Book of Genesis, seems to have originated with the Jews. In picking the number seven, they may have been influenced by its supposed mystical properties (seven was considered a lucky number even in ancient times). The fact that seven days is roughly one quarter of a lunar month (that is, it approximates the interval from new moon to half-moon, or from half-moon to full moon) may also have impressed them. The effect of the adoption of the week was to add still another unit to the calendar, one that was out of step with both the month and the year.

Even with such cumbersome additions, the Julian calendar was much simpler than the Babylonian calendar, and almost as accurate. Over the centuries, however, a built-in error—a gain of one day every 128

CALENDAR REFORMER Christopher Clavius was an astronomer-mathematician and an authority on sundials. A member of the papal commission on calendar reform, Clavius wrote two closely reasoned Latin tomes defending the new Gregorian calendar after it had been approved by the Pope. His arguments convinced many doubters of the calendar's worth and led to its eventual acceptance throughout most of the world.

years—piled up. By the 16th Century, the calendar was running 13 days behind the sun. This discrepancy had little effect on the lives of ordinary folk, but it disturbed the Church, because it pushed holy days into the wrong season (Easter, for example, was well on the way to becoming a summer festival).

In 1582 Pope Gregory XIII, after long consultations with the Italian physician and astronomer, Aloisius Lilius, and a German mathematician, the Jesuit Christopher Clavius, decreed that the following year should be shortened by 10 days. Gregory also ordered a revision of the leap-year system of correction, calling for the omission of three leap years in every four centuries. As a result, the Gregorian calendar is accurate to within one day in every 3,323 years.

Adoption of the modern calendar

Gregory's innovations, of course, were immediately adopted in Catholic countries. The Protestant lands, however, had only recently gone through the Reformation and were in no mood to accept Roman dogma, even on so neutral a subject as the calendar. They clung to the old Julian calendar despite its errors. Gradually, however, the merchants and diplomats in these countries became aware of the inconveniences of using two calendars—the Old Style and the New Style—on one small continent, and one by one the Protestant lands swung into line with the more rational system. England, always strong on tradition, was the last to yield, holding out until 1752, by which time it was 11 days out of step with its neighbors.

Once England (and its American Colonies) had adopted the New Style, the only important European holdout was Russia, whose Orthodox Church had broken with Rome long before the Reformation. Russia kept the Julian calendar—whose errors, of course, continued to accumulate—for nearly two centuries more. In 1918, as one of the minor results of the Bolshevik Revolution, the Russian government dropped 13 days from the year to bring its calendar in line with that of the rest of Europe. As a result, the annual Soviet celebration of the "October Revolution" (marking the Communist takeover in Russia), now comes on November 7.

The Orthodox Church has never accepted the godless Bolsheviks' reform. To this day it retains the Julian calendar and celebrates Christmas (by the Gregorian reckoning) on January 7. The Orthodox Christians are no more tenacious in such matters than the Orthodox Jews. Today, as for more than 25 centuries, the Jews reckon their holy days by the lunisolar calendar which they acquired by the waters of Babylon.

Perhaps the strangest calendar in wide use today is that of the Moslems. The caliph Omar, who followed Mohammed as the leader of Islam, took over the lunisolar calendar, which was in general use throughout the Near East. For some reason he threw out the system by which extra months were occasionally added to keep the calendar in phase with the seasons. As a result, the Moslem calendar today is purely lunar; its year consists of six 29-day months and six 30-day months, for a total of 354 days. The seasons and the months have no connection, and 33 of their years come to about 32 of ours. In the Moslem world the calendar is used to schedule religious feasts; and consequently these feasts shift through the year. To avoid confusion, Moslem countries use the Western calendar to schedule their civil events.

Our own civil calendar, though simple and accurate enough for our needs, has been criticized for its irrational divisions. The months are not of the same length, nor are the quarters, which many businesses use for accounting and planning purposes. Neither the months (except February) nor the year contain a whole number of weeks. As a result, every year and 11 out of the 12 months begin on a different day of the week from the preceding year or month.

The proposed world calendar

To remedy these deficiencies, reformers for more than a century have proposed various methods of revising the divisions of the Gregorian calendar, while leaving intact its accurate method of handling leap year. Perhaps the most widely accepted of these proposals for calendar reform is the World Calendar.

This calendar, which was devised in 1930 by Elizabeth Achelis, a calendar-reform enthusiast, divides the year into four equal quarters, each of 91 days. Each quarter, in turn, is divided into three months, of 31, 30 and 30 days. The quarters contain precisely 13 weeks, and thus each three-month period of this relatively recent World Calendar begins on a Sunday and ends on a Saturday.

Since four quarters of 91 days add up to only 364 days, an extra day is added to the month of December. This day is dated December W and is completely outside the week. Though it falls immediately after a Saturday, it is not called Sunday, but Worldsday. In leap years another "nonweek" day, which is called Leap-Year Day, is added at the end of June.

The World Calendar has been discussed by the United Nations Economic and Social Council and has won favorable comments from a number of national representatives there. But religion, which has been so centrally involved in time-reckoning from the beginning, is not yet ready to relinquish its role. While the major religious denominations of the world have raised no objections to the World Calendar, it has drawn the fire of several minority groups, notably the Orthodox Jews and the (Protestant) Seventh Day Adventists. These groups share a literal interpretation of the Bible's injunction: "The seventh day ... thou shalt not do any work." They believe that Worldsday would disrupt the regular, seven-day cycle envisaged by this commandment.

These religious objections, combined with human inertia, are likely to block adoption of the World Calendar for many years, perhaps indefinitely. For the foreseeable future, the calendar on the kitchen wall is likely to remain illogical and unchanged.

THE WORLD CALENDAR shown here would reform the Gregorian reckoning by assigning permanent dates to the days of the week. On this calendar January 1 always falls on Sunday, and other holidays except Easter come on the same day every year. This is accomplished by dividing the year into equal quarters, each containing one month of 31 days *(blue)* followed by two of 30. As the 12 months would add up to only 364 days, an extra day—Worldsday—is placed between the last of December and the first of January, but is not counted as part of either month; in leap years a second Worldsday, belonging to no month, is added between June and July. This system has been pushed by the International World Calendar Assocation but its acceptance is doubtful.

The Riddle
of the Calendar

"Thirty days hath September ... " begins the old rhyme recited by generations of schoolchildren. Occasionally a youngster chanting the lines may stop, puzzled by the ragged distribution of days: 30 for four of the months, 31 for seven, normally 28 for February, "and 29 in each leap year." This same problem has tantalized priests and popes, astronomers and astrologers, who have struggled with its larger aspects: how to make the days and months fit into the year. Behind the effort lies a dilemma: The year, which determines the seasons, is based upon one revolution of the earth around the sun, but that time comes to more than 12 new moons, or months. Precisely stated, a year lasts 365 days, five hours, 48 minutes and 46 seconds. But the three astronomical cycles establishing the year, the month and the day are independent and incompatible: Like imperfectly matched gears, they do not mesh. Men have tried gamely to divide the year in such a way that important days—holidays, vacations, ceremonies—will be in tune with the seasons year after year, century after century. Ingenious calendars have been devised, but a fully accurate solution cannot be found, because the problem of reconciling the days, the months and the year is really insoluble.

THE ENDLESS ROUND OF TIME
A calendar devised by the Portuguese cartographer Diogo Homem in 1559 lists the months along its outer rings and under them the time of appearance of each new moon throughout a 19-year cycle. Thus the calendar is a perpetual one, never out of date. In the corners are other dates, such as those for phases of the moon and religious holidays.

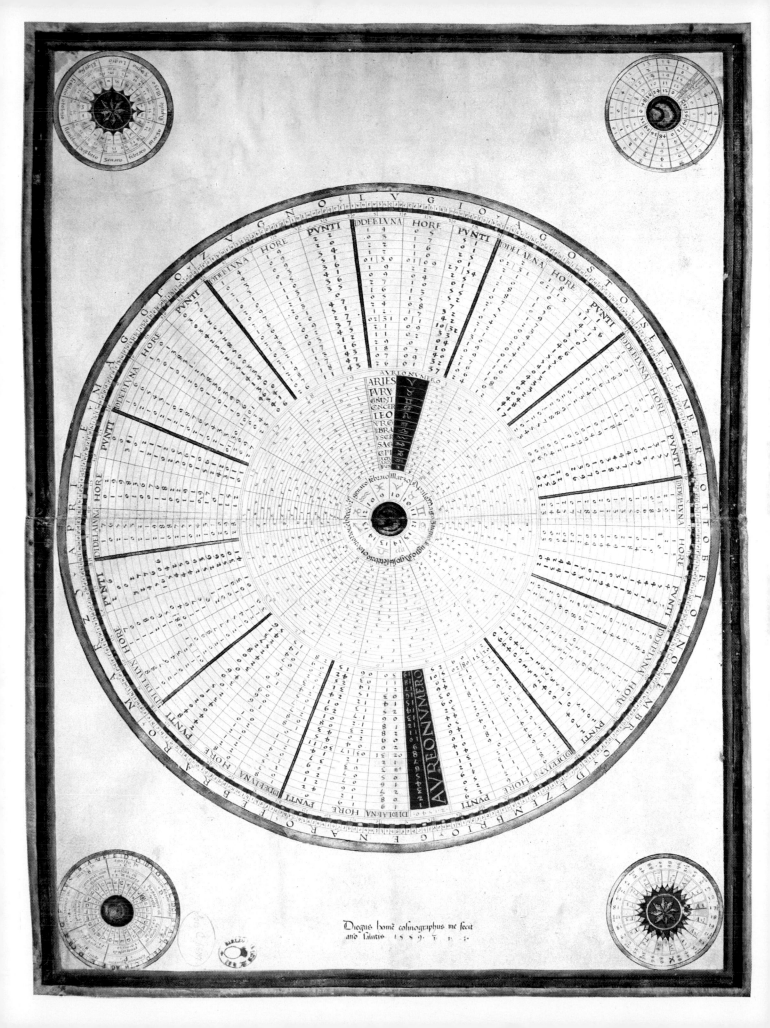

Diegus homē cosmographus me fecit
año Calutis 1589

Phases of the moon—from new moon *(left)* to new moon *(right)*—fixed the month on which the Babylonians based their calendar.

Fixing the Months
by the Lunar Cycle

Even in prehistoric times man had to keep track of the passing days so that he could tell when to hunt migrating animals or seek shelter for the coming winter. But not until the complex civilizations of the Tigris-Euphrates Valley emerged did the need to know when to plant crops or prepare for floodtime give rise to the first true calendars. By 2000 B.C. the Babylonians had devised a calendar based on the average period of 29½ days between new moons. In it the year was divided into 12 lunations, or months, for a total of 354 days. Since this count fell short of the solar year by 11 days, it was not long before harvest rites were taking place in the wrong seasons. To ensure a proper relationship of ceremony to season, the priests hit upon a device still in use — intercalation, the addition of extra days or months to correct the mismatched astronomical cycles and thus bring the calendar into harmony with nature. At first, months were added at the whim of priests, but later a schedule was set in which seven additional months were spread over a 19-year cycle to bring the months and years in phase.

The Babylonian system was a model for Hebrews and Moslems but each made one great change: The Jews introduced the seven-day week—a unit roughly equal to a quarter lunation; the Moslems discarded the corrections to establish the only purely lunar calendar still in use today.

A Babylonian calendrical tablet records, from left to right in lines of cuneiform script, the intervals from new moon to new moon and the

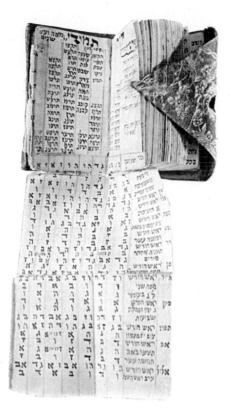

A POCKET HEBRAIC CALENDAR
A Hebrew calendar records Jewish festivals for the year. Though it dates from 19th Century Poland, it embodies the 4,000-year-old Babylonian calendar system: 12 lunar months, corrected by adding seven months every 19 years.

lunar positions during a 25-month period for the years 103 to 101 B.C.

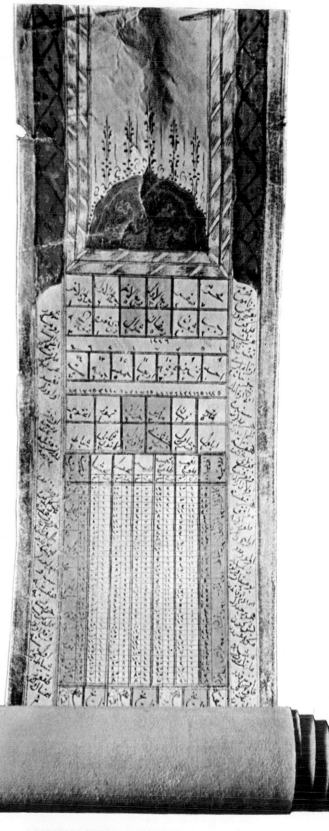

A MOSLEM SCROLL CALENDAR
In this calendar dating from the Islamic year 1225 (1810 A.D.), the months and weeks are listed in horizontal rows near the top. The vertical columns beneath tell the exact times of dawn, sunrise, afternoon prayer hour and sunset.

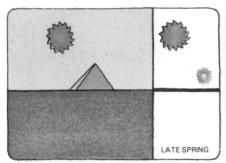

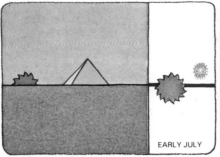

LATE SPRING

EARLY JULY

JULY 19

A STARRY NEW YEAR

The Egyptians marked their new year by the first visible rising in the east of Sirius, the "dog star." The drawings above are in two parts, the section at left showing what was visible to the Egyp- tians, at right indicating the actual positions of Sirius and the sun. In late spring, when Sirius rises, it is still obscured by the sun. In early July *(center)*, the sun is just below the horizon when Sirius rises, but dawn light blanks out the star. However, about July 19 *(right)*, when the sun has fallen farther behind Sirius, the star's presence could at last be perceived.

In this Egyptian calendrical table from the tomb of Ramses VI, the sky goddess Nut *(face at far right)* lies stretched out between the

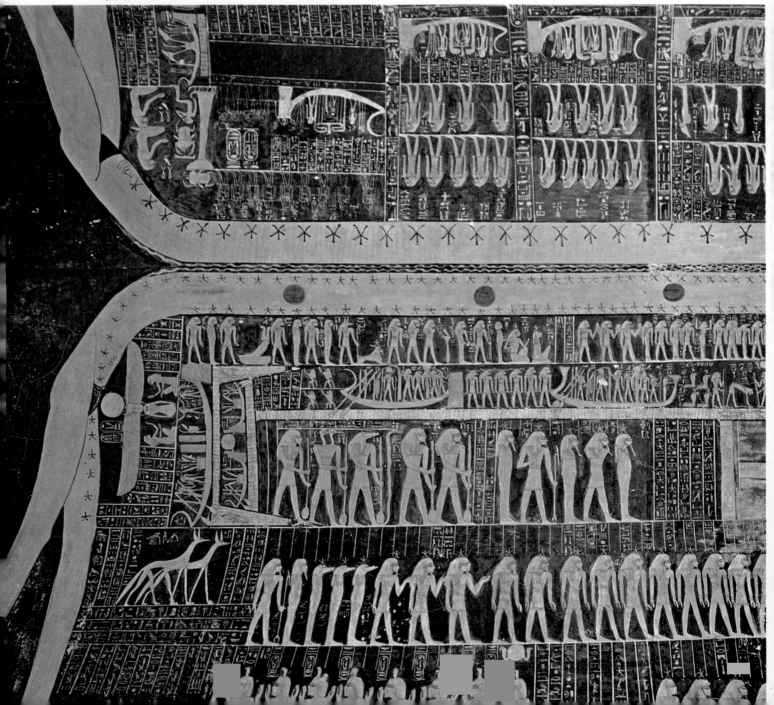

Egypt's Legacy: The 365-Day Year

Late spring in the Valley of the Nile was the time when Egyptians gazed skyward and waited for the vitally important flood that would begin soon after the appearance of the star Sirius; with this event the Egyptians started their new year. They also kept a separate year made up of 12 fixed 30-day months to approximate the lunar cycle. Later, to make their lunar year jibe almost precisely with Sirius' rising, they tacked five extra days onto the year. To account for them they created the myth of Nut, the sky goddess, who had been unfaithful to her husband, Re, the sun god. In retribution, Re decreed that she should bear a child "in no month of no year." But Nut's lover Thoth played dice with the moon and won five days a year. Because these days were outside the calendar, Re's decree did not apply. Nut's son was born on the first of them.

Mythology aside, the Egyptians did give us the 365-day year, and later their astronomers unsuccessfully proposed that the year be brought into closer phase with the sun by adding an extra day every fourth year.

deities of the day *(bottom panel)* and night. The table was one of several that interpreted time's progress in mythological terms.

Ancient Maya Monument to Time

Literally worlds apart from the European tradition was the amazingly accurate timekeeping of the Central American Maya. While other cultures were concerned with time, the Maya were obsessed with it. Their calendar was a part of their religion; and the task of synchronizing it with nature fell to the priest-astronomers, who used three great astronomical cycles—the daily rotation of the earth, the lunar month and the solar year. They did not attempt to reconcile these cycles but kept separate records of these three, as well as other astronomical cycles.

At frequent intervals the priests checked and matched the separate calendrical notations for religious holidays all the way back to the beginning of time (which they set at 3113 B.C.). Then they inscribed the corrected dates in almanacs or on tall stone shafts, called stelae. As a result of this careful cross-checking, the Maya calendar was more precise than ours. Such precision was essential because, in the words of a leading scholar, the Maya "conceived of the divisions of time as burdens carried through all eternity by relays of divine bearers." Every day, year, decade, century and millennium had its own bearer-gods. Only by knowing which gods would be bearing the burdens could the Maya know whom to propitiate. The priests had access to copious records of past courses of the planets (particularly Venus), the sun, the moon and the stars, all of which were correlated with the activities of the gods. From these past celestial movements, priests could make accurate predictions and thus determine which gods would reign at a given time.

782 A.D., MAYA STYLE

At Copán, Honduras, site of the great Maya city, a nine-foot (2.7-m) stela honors an astronomer-priest and marks a time period corresponding to 782 A.D. Hieroglyphs on the stela record the date, the phase of the moon and the name of the god who then held sway.

THE MEANING OF THE GLYPHS

In the hieroglyphic system used by Maya priests to fit dates into the calendar, numerical symbols were joined with pictures of the gods—much as if we were to write March 13 as 13 plus a drawing of Mars, god of war. Thus three dots with a glyph representing the god Etznab *(above, left)* meant the third day of the period ruled by Etznab; two bars and two dots with Lamat meant the 12th day of Lamat's rule. Both these glyphs, shown in the detailed drawing above are indicated by arrows in the Maya almanac reproduced below.

A RECORD OF THE VENUS CYCLE

A page from a 12th Century A.D. Maya almanac lists the dates when Venus and the sun rose together, as well as the disappearances and reappearances of the planet. The arrows indicate such a heliacal rising for the 12th day of Lamat and the appearance of Venus as the evening star on the third of Etznab.

Julius Caesar's Calendar Reform

For a thousand years—from the beginning of the Christian era through medieval times—the Julian calendar was used throughout the Western world. Named after Julius Caesar, the great Roman who instituted it, the calendar was a great improvement over its predecessor, a haphazard timekeeping arrangement, often manipulated for political ends.

In Rome's early days its calendar, linked to the moon, had 10 months. As these added up to about 300 days, extra days were added as needed to keep time in phase with the seasons. Although the year was divided into 12 lunar months in the Eighth Century B.C., the names of the last four months of the old calendar survived —September, October, November and December, the Latin numerical designations for the seventh, eighth, ninth and tenth months.

When Julius Caesar returned from Egypt in 47 B.C. he established a year of 365 days adjusted by leap years, as Egyptian reformers had suggested two centuries earlier. Now the calen-

SPRING: A TIME FOR PLOWING
The Duc de Berry's calendar, like so many today, combined a modest amount of information with pretty pictures. In March a peasant is plowing; concentric arcs give the dates of the new moon for 19 years. On top are the zodiac signs Pisces and Aries.

SUMMER: A TIME FOR HAYING
Peasants mow hay with scythes in June while their wives gather it into stacks. Beyond the verdant fields rises the royal palace of King Charles VI in Paris. The sun god, Phoebus, rides his chariot above the picture. The zodiac signs are Gemini and Cancer.

dar began to approach modern accuracy and form. The months, though not exactly coinciding with ours, had 30 or 31 days; then as now, February was the exception, with 29 days ordinarily and 30 in leap years. While Caesar's successor Augustus made some adjustments—such as cutting a day from February—it was not until 527 A.D. that major changes were made. Dionysius Exiguus, the abbot of Rome, moved New Year's from January 1 to March 25—perhaps to reflect nature's annual rebirth. The abbot also fixed Christmas at December 25 and began the practice of dating events (B.C. and A.D.) from the birth of Christ.

It was not until the Middle Ages that laymen became interested in calendars. Aristocrats ordered richly decorated books of hours for their enjoyment—and as a convenient way to schedule their activities. Perhaps the most beautiful of these ornate calendars was the one commissioned by the Duc de Berry in 1409, portions of which are reproduced below.

AUTUMN: A TIME FOR HARVESTING
In September grapes are gathered in fields around the Château de Saumur. Here the signs of the zodiac are Virgo and Libra. The Duc de Berry, a great patron of the arts, commissioned the Flemish painter Pol de Limbourg to illustrate this book of hours.

WINTER: A TIME FOR WARMING
On a cold February day a red-capped farmer and his family warm themselves before a fire, while outdoors a peasant chops a tree and another prods a donkey laden with firewood. The zodiac signs are Aquarius, the water bearer, and Pisces, the fishes.

Pope Gregory's Finishing Touch

The Julian calendar's simple form enabled a fairly wide cross section of people to keep track of time and order their affairs. But despite its leap years, it failed to mesh with the astronomical cycles closely enough: The average calendar year was 12 minutes longer than the solar cycle. This seemingly trivial error accumulated so that in 1093, for example, spring fell on March 15 instead of March 21, and movable feasts, like Easter, slowly slipped back through the seasons.

In 1582, Pope Gregory XIII promulgated a new reform. Centurial years not divisible by 400—1700, 1800, etc.—were no longer to be leap years. This reduced the annual error to only 26 seconds, which comes to one day every 3,323 years. To bring the calendar back into step with the seasons, Gregory reduced the year 1582 by 10 days, October 4 being followed by October 15. As a final fillip he restored January 1 as New Year's Day. Catholic Europe adopted Gregory's new calendar but the Protestant states held back. Only in 1752 did England and its Colonies conform by cutting 11 days from the year. This event touched off riots in London, where many, angry because they had been cheated out of 11 days' rent money, rioted to the cry, "Give us back our 11 days." In America, however, Ben Franklin counseled a more philosophic view, telling his readers not to "regret . . . the loss of so much time" but rather to rejoice that one might "lie down in Peace on the second of this month and not . . . awake till the morning of the 14th."

PRESENTING THE PAPAL REFORM
Pope Gregory XIII is shown in this 16th Century painting presiding over a meeting in Rome to present his calendar reform. A savant, pointing to zodiacal signs associated with the months, explains the innovation to the assemblage while priests and laymen discuss the new calendar.

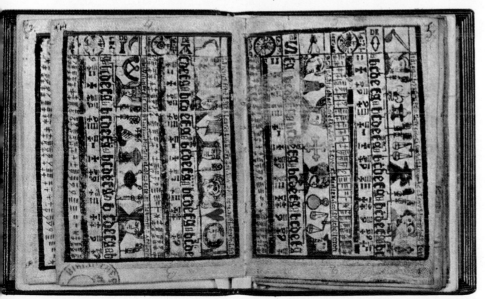

A MARINER'S CALENDAR
Two pages from a 17th Century perpetual calendar used by Breton seamen show pictures of the saints *(right column)* alongside their days and then in red a listing of the days of the lunar month. Other portions of the calendar included maps and tide readings for the European coast.

4
Ticking Off the Hours

Nicholas Dratzer, 16th Century horologer to Henry VIII of England, works on a sundial which has 10 faces for telling time in different areas. Instruments like sundials which told time by shadows are the oldest known clocks, dating back to 3000 B.C.

WITHOUT MEASURES FOR TIME our highly organized civilization would grind to a halt. Hours, minutes and seconds are woven into its very fabric. Every morning millions of commuters drive to stations in cars rated in horsepower (foot-pounds per second), listening to "on the hour" broadcasts and then sprint down platforms to catch trains that leave— or at least are supposed to leave—on the minute.

Time is also one of the fundamental quantities of nature, and time measurement is at the foundation of science. The units are written into its basic vocabulary. The astronomer deals in light-years and Julian days; the geophysicist measures the acceleration of gravity in terms of centimeters per second per second; the pharmacologist must determine how many hours a drug remains in the body.

Time measurement, of course, involves clocks, but clocks are only a part of the story. Before time can be measured, there must be a concept of time—the subtle notion that there is something to be measured—and before this notion can be acted on, there must be time units, small units more serviceable than the "natural units" of day, month and year.

The early history of small time units parallels that of the calendar. The Sumerians, who were the first to divide the year into units, were also the first to divide the day, and they followed the same pattern. Just as their year contained 12 months of 30 days each, so their day consisted of 12 *danna*, each of 30 *ges*.

It is not the Sumerians, however, but the Egyptians to whom we owe the 24-hour day, just as we do the 365-day year. Their reason for establishing a system of hours was almost certainly religious. The Egyptian word for "hour," *wnwt*, is also the word for "priestly duties"; with one added hieroglyph it becomes *wnwty*, meaning both "hour-watcher" and "star-watcher." The star-watchers carried out their priestly duties by noting the appearance of the "decans"—particular stars or constellations—on the eastern horizon. They divided the night into 12 hours, and each hour was marked by the appearance of an appropriate decan.

The daylight hours, of course, were marked by the heavenly progress of the sun-god Re. To keep track of the sun, the priests used shadow clocks. One version was shaped like a T square but with the crossbar of the T turned up at right angles. When the instrument was pointed east (in the morning) or west (in the afternoon), the shadow of the crossbar fell across the handle and registered, rather inaccurately, the passing hours. To number the hours of the day, the priests simply used the numbers from one to 10. They reckoned the "twilight hours" separately, one hour for dawn and another for dusk. These hours plus the 10-hours day and the 12-hour night made up the full complement of 24.

Because of the way this system was set up, the Egyptian 24-hour day differed markedly from ours: Its hours were not the same length. The daylight hours, each one tenth of the interval from sunrise to sunset, were necessarily longer in summer than in winter. The night and twilight hours varied in a different and more complicated way, as the rising of the decans shifted from season to season and from year to year. Even for the Egyptians, who cared but little for consistency, this system was too complicated. Eventually they decided that both the day and the night should be reckoned as 12 hours long, eliminating the separate hours for dawn and dusk. The hours still varied in length from season to season but at least the variations were consistent.

The watery flow of time

This simplification of timekeeping may have been connected with the invention of the first clocks that were not directly dependent on the heavens. These were water clocks, or clepsydras, which measured time by the emptying or filling of a vessel, as water flowed out of it or into it through a small opening.

The clepsydra made possible a new way of looking at time. Sun and star clocks alike were primarily indicators, marking off the hours at which certain activities were scheduled to be performed. But the clepsydra, by the emptying or filling of its vessel, also gave clear and concrete evidence of *how much* time had passed. In this sense, its invention can be called the real beginning of time measurement.

After the clepsydra, the next advance in timekeeping instruments came from the Greeks of Alexandria. In the centuries just before the Christian era, they applied their mechanical and scientific genius to improving the water clock and the sun clock (which by now had become the sundial). For both the clepsydra and the dial, however, they continued to use the old Egyptian scheme of seasonally variable hours.

As timekeeping devices improved, clocks outgrew their essentially religious functions. In the classical world they were used for a variety of secular purposes. Attorneys in the law courts of Rome, for example, were allowed a certain amount of clepsydra water for their speeches. Unscrupulous lawyers sometimes bribed a court attendant to fill the clock with muddy water which would clog the outlet, thus allowing them to "extend their remarks." Clock time — measured time — was becoming part of the lives of ordinary people. Gradually, as the clepsydra became more familiar, so, too, did the notion of time as a thing in itself, as a flowing reality that could be measured independently of the heavens.

The classical world had, as it were, domesticated time, and yet time still could not be measured in any very accurate way, since the measuring rod — the hour — expanded and contracted with the seasons. The first important step toward timekeeping with fixed time units was taken by Moslem scientists of the early Middle Ages. The Moslems may have inherited the notion of fixed hours from the Babylonians, whom some authorities credit with a day containing 12 "hours" of equal length. Whether this is true or not, we know that around the beginning of the 10th Century a Moslem astronomer made a much improved type of sundial with only a single scale. This new dial could mark off the hours accurately the year round — provided the hours were of uniform length.

What finally put uniform hours into general use, however, was the invention soon after 1300 A.D. of the mechanical clock, which reckoned

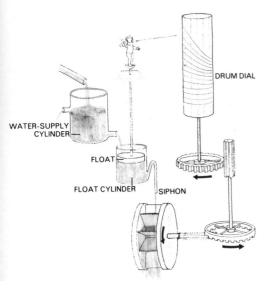

A GREEK WATER CLOCK, shown here in a baroque reconstruction, kept track of the hours and days in a whole year. Water flowed into a cylinder *(upper left)* which had a small overflow to keep its water level, thus maintaining a constant pressure and a steady outflow from its bottom. This flow filled a second cylinder, raising a float that carried an angel; the angel indicated the hour on a drum dial. The dial compensated for the ancient idea that there were 12 hours between sunrise and sunset regardless of the time of year; the dial's hour lines narrowed so that as winter approached, the span of the hours shortened. The drum moved one day at a time; the movement occurred when a narrow siphon on the float cylinder emptied water onto a waterwheel and turned a series of gears.

WATER-SUPPLY CYLINDER

DRUM DIAL

FLOAT

FLOAT CYLINDER   SIPHON

time by counting uniform, periodic motions — a principle which is applied even in the superaccurate atomic clocks used by scientists today. Curiously, the origins of this major step in time measurement were until recently almost unknown.

Even the crudest mechanical clocks were quite elaborate structures, involving the interaction of many parts. They were powered by a weight fastened to a cord wound around an axle, or "barrel"; as the weight lowered, the cord forced the barrel to revolve. The barrel, in turning, set in motion a train of gears, which controlled bells for striking the hours and a pointer for indicating them. The most ingenious part was the escapement, a device whose oscillating motion periodically stopped the clockwork, compelling it to revolve at a slow, steady gait.

For years historians wondered how so elaborate a mechanism could have appeared seemingly out of the blue. The weight drive, to be sure, was no problem: Even the Romans knew of the windlass, which raises weights by winding a cord around an axle, and the weight drive is a windlass in reverse. It was, in fact, used by medieval Moslems for hoisting water. But where did the gears and the escapement come from?

The case of the clock's missing pedigree

The answers did not appear until 1955, when Professor Derek J. de Solla Price, then at Cambridge University, set out to track down the mechanical clock's pedigree. The clockwork proved a fairly easy problem. Price turned up a whole family tree of "geared, clock-work-like devices . . . before the advent of the clock." Some were "planetariums," mechanical models which displayed the motions of the heavens; others were crude machines for computing lunar and planetary motions.

The search for the origins of the escapement proved more difficult, however. But at Cambridge one day, while seeking information on a Chinese planetarium, Price consulted Professor Joseph Needham, probably the greatest Western authority on the history of Chinese science and technology. When the two men examined the relevant historical sources, they spotted an illustration of "an intriguing arrangement of rods, pivoted bars and levers that seemed . . . to act as an escapement, checking the motions of the wheels." Careful translation of the text established that the arrangement was indeed an escapement and that the planetarium also served as a clock.

The remarkable device was built in 1088 A.D. by a Chinese mandarin named Su Sung. Further research established that the first escapement clock had been built more than 300 years earlier. Thus it was Chinese clocks, contemporary with Europe's Dark Ages, that bridged the gap between the clepsydra and the mechanical clock. There is, to be sure, no direct evidence that word of these clocks ever reached Europe. But Price, at least, strongly suspects that it did and that it stimulated Europeans to build similar devices. A clever man, once he knows something has been done, can often work out his own way of doing it. This does not mean, of course, that the first European escapements were merely duplicates of the Chinese clocks. There were important differences. Instead of "pivoted bars and levers," the European clocks employed an equally elaborate device called the verge and crown wheel. This mechanism alternately blocked and unblocked the motion of the clockwork, forcing it to turn at a measured pace.

But though the pace was measured it was far from accurate. The difficulty was basic to the verge escapement: It had no built-in "time sense" of its own. Its blocking and unblocking motions were not isochronous, or evenly spaced, because they depended on the interaction of all the parts—every one of which was subject to the vagaries of handcraftmanship and wear. As a result, the first clocks did well if they gained or lost no more than 15 minutes a day. But accurate or not, the hours which they struck off were hours of fixed length. To preserve the old scheme of variable hours would have meant rearranging the mechanism every few weeks, a task that no clockmaker of the time cared to undertake.

After the introduction of the verge escapement, no basic improvement in clocks appeared for some two centuries, though better workmanship and design made them more accurate. The invention of the more compact spring-driven clock, probably some time in the 15th Century, provided portable timepieces. But these ancestral watches were even less accurate than the weight-driven clock.

The pendulum's perfect period

The next big step toward improved accuracy was the handiwork of the great Italian scientist Galileo, who discovered a source of evenly measured (or at least nearly evenly measured) oscillation. The source was a familiar one: a swinging weight, or pendulum. While watching the movements of a hanging church lamp, the story goes, Galileo noticed that it seemed always to take the same amount of time for one complete swing; that is, its period of oscillation was constant. This led him to deduce that the period of a pendulum depended on its length, not on the magnitude of its swing or even on its weight. (As we now know, he was only approximately correct.) Toward the end of his life, he even devised an escapement by which a pendulum could regulate a clock or some similar device but, aging and almost blind, he never built it.

The man who did build the first workable pendulum clock, some 14 years after Galileo's death, was another great astronomer, the Dutchman Christian Huygens. His mechanism, far more accurate than any existing clock, proved an immediate commercial success. "It is a strange thing," Huygens dryly wrote to a friend, "that before me nobody spoke of such clocks, yet now so many other inventors have turned up."

The fact that such a distinguished scientist as Huygens spent time designing better clocks is evidence of another aspect of time measurement: its growing importance to science. Galileo used a pendulum to time astronomical observations and a water clock to study moving bodies. Even earlier, improved clocks helped refine astronomical observation, thereby providing the data from which, a generation or so later, the German

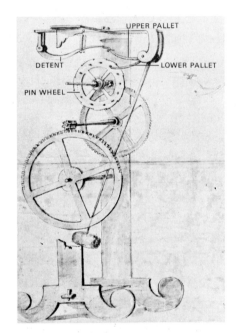

GALILEO'S CLOCK MECHANISM, seen in this 1641 drawing by his son Vincenzio, incorporated the pendulum as its regulator, along with a "pin wheel" escapement. As the pendulum swings to the left, the top pallet lifts the clawlike detent and lets the pin wheel go forward. The bottom pallet then swings up and halts the wheel by engaging a pin on its side. In this way, the detent and bottom pallet, working in alternation, keep the pin wheel and clock gears turning in regularly spaced steps.

astronomer Johannes Kepler deduced his historic laws of planetary motion.

These and many other research projects produced a steady demand for ever more accurate clocks, which scientists and clockmakers were quick to supply. They worked out ways of making the pendulum more even and accurate in its motions and cleverly redesigned it so that its length—and therefore its period—would not change with changing temperature. They also tinkered with the escapement by which the pendulum controlled the clockwork's movement. Less than three generations after the first pendulum clock began ticking, these improvements had combined to produce the regulators, highly accurate pendulum clocks which gained or lost no more than a few seconds a week and were therefore especially suited for scientific use.

Ships that navigate by the clock

Even more avid than the scientists in their search for more accurate clocks were sea captains. For them, precise timekeepers could mean the difference between life or death, because without such reliable instruments they could not know the location of their ships.

The geographical position of a ship at sea (or anything else, for that matter) is of course determined by its latitude and longitude. Latitude poses no problem. As far back as ancient Greece, it was known that the elevations of the sun and fixed stars change, not only with the seasons, but also as the observer travels north or south, to a higher or lower latitude. To find his latitude, the navigator simply measures the altitude, or angle that a fixed star such as Polaris makes with the horizon. By the 17th Century at the latest any ship's officer worth his salt could measure his latitude to a fraction of a degree.

Longitude, however, was another story. Seamen and scientists alike had agonized for centuries over the problem of determining the longitude. Many methods were proposed, but all were impractical, if not outright fantastic. Meanwhile, mariners continued to estimate their longitude by dead reckoning—which means partly by guesswork—and to run their ships ashore when they guessed wrong.

For Britain, the last straw came in 1707, when a fleet commanded by an admiral with the comic-opera name of Sir Cloudsley Shovel mistook its longitude and ran into the Scilly Islands. The loss was four ships and 2,000 men, including Sir Cloudsley. After a lengthy study of the problem, the government in 1714 offered a prize of £20,000 (a small fortune in those days) for a method of determining the longitude to within half a degree.

There had always been—in theory, at least—a method for finding longitude. Like the determination of latitude, it depends partly on observation of the sun or stars. Their relative positions do not change with longitude, that is, with east-west travel; they are precisely the same in London and at St. Lunaire Bay, off the coast of Newfoundland, which lies in the same latitude but is some 55°20′ of longitude to the west. Still there is an east-west difference in the sun and stars. It is one of time: Off the coast of New-

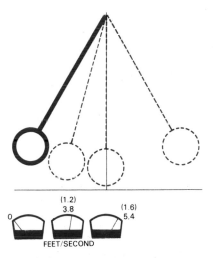

THE PENDULUM was for centuries the most accurate clock regulator because the time it takes to complete one swing stays nearly constant despite changes in the length of the swing. In the sketch above, a pendulum's swing is timed through half of a 60° arc. As the speedometer beneath each position shows, the pendulum accelerates from a dead stop at the top of its arc to 3.8 feet (1.2 m) per second halfway down and to 5.4 feet (1.6 m) per second at the bottom. Later (below), friction has reduced the arc to 30°, but the speed is also reduced by half. So in both cases the arcs are completed in the same total time interval: one second. Meters per second are in parentheses.

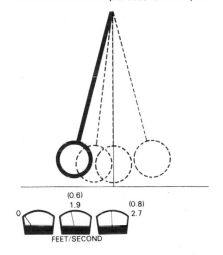

foundland they rise three hours 41 minutes later than at London. A mariner can determine how far west of London (and therefore how far east of Newfoundland) his ship lies by comparing the local time—sunrise, for example—at his position with the time at that same instant at London. If he finds sunrise comes just three hours later than it did at London, he knows that since sunrise at London the earth has turned one eighth of a day, or 45° of a 360° rotation. He must therefore be 45° west of London and about 10° east of St. Lunaire Bay. But he can compute all this *only if he always knows what time it is at London.*

The obvious way to keep track of the time at London is to carry a clock set to London time. What was needed to solve the longitude problem, said Sir Isaac Newton, one of the British government's scientific consultants, was "a Watch to keep time exactly." But, he added, "by reason of the motion of a Ship, the Variation of Heat and Cold, Wet and Dry, and the Difference of Gravity in different Latitudes, such a Watch hath not yet been made."

The first man to make "such a Watch" was John Harrison, son of a Yorkshire carpenter. By 1728, Harrison had made drawings of a marine timekeeper and six years later he submitted his first clock—a monster weighing 72 pounds (32 kg)—to the Board of Longitude, which evaluated proposed navigational methods. A preliminary trial at sea gave encouraging results, and the Board voted him £500 to produce an improved model.

Over the next quarter of a century, Harrison built three more clocks; his last was no bigger than a modern alarm clock. In 1761, he was ready for a trial. On November 18, H.M.S. *Deptford* sailed from England for Jamaica, carrying Harrison's "No. 4," in charge of his son, William.

The clock that slaked a crew's thirst

The first stop was to be the island of Madeira. After nine days out of sight of land, the ship's longitude by dead reckoning was estimated at 13°50′ west. William, however, using No. 4, calculated it at 15°19′, and predicted that if the ship held her course Madeira would be in sight the following day. The next morning, at 6 a.m., the lookout sighted the nearby island of Porto Santo. Had the ship changed course in conformity with the dead reckoning longitude, she would have missed Madeira entirely, "the consequence of which," says a contemporary account, "would have been inconvenient, as they were in Want of Beer."

With the missing beer replaced by a dozen casks of Madeira wine, the *Deptford* pressed on to Jamaica. On arrival there, the longitude determined by the timekeeper proved accurate, not only to within half a degree, as the prize specified, but to within about 1/50 of a degree.

The drama attending Harrison's achievement has obscured an ironic fact: Though his timekeeper was the first to measure time accurately at sea, it exerted little influence on future chronometers. The true "father of the chronometer" is generally conceded to be the Frenchman Pierre LeRoy. Working independently of Harrison, LeRoy in 1766 produced a timekeeper of radically different design, especially the escapement.

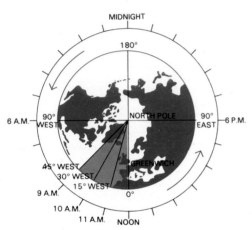

HOW TIME MEASURES DISTANCE is illustrated in this diagram, which shows a 24-hour clock dial surrounding a map of the earth as seen from the North Pole. The earth goes through one rotation of 360° every 24 hours; thus each hour on the clock corresponds to 15° of distance *(shaded sectors)*. The starting point for measuring longitude is Greenwich, England; when it is noon there, the time at a point 15° west will be 11 a.m., at 30° west it will be 10 a.m., at 45° west 9 a.m., and so on around the globe.

He separated it from the drive mechanism to reduce friction, built in an automatic compensation for temperature variations and provided an easy adjustment—changes which improved precision while simplifying construction. When tested, his clock proved as accurate as—even though of much cruder workmanship than—Harrison's No. 4.

Defining the second with precision

With the emergence of instruments that could measure time accurately to within a few seconds on land or sea, it was natural that the accuracy of time units themselves should be questioned, that somebody should ask, "Just how long is a second?" The question was not so silly as it sounds. Simple mathematics makes it clear that a second is 1/60 of a minute, which is 1/60 of an hour, which is 1/24 of a day—that is, one second equals 1/86,400 of a day. But in fact, because of variations in the earth's orbital speed and its distance from the sun, a solar day—the interval from noon to noon—is not the same length all year round.

This imprecision was first questioned seriously during the French Revolution, when the disorder among measuring units was straightened out by introduction of the metric system. In 1820 a committee of French scientists recommended that day-lengths throughout the year be averaged and that a second be defined as 1/86,400 of this mean solar day. The new definition, which was used for more than a century in most countries, supplied science with an internationally accepted standard of time.

Time intervals were thereby standardized—but not time epochs. Each locality still ran on its own time. Clocks in each town and city were set to mean solar noon—that is, to sundial noon corrected to allow for variations in the length of the solar day. With the coming of the railroads, these easygoing local arrangements had to end. The more communities a railroad operated in, the more different times it—and its shippers and passengers—had to cope with.

At first railroads used their own "standard" time, but this compounded the problem. Around 1880, the railroad station in Buffalo, New York, displayed three large clocks, one set to local Buffalo time, another to New York City time, favored by the New York Central railroad, the third to Columbus, Ohio, time preferred by the Michigan Southern.

On October 11, 1883, a General Time Convention of the railroads divided the United States into four time zones, each of which would observe uniform time, with a difference of precisely one hour from one zone to another. With certain modifications, these zones still define the areas of today's familiar Eastern, Central, Mountain and Pacific Standard Time. In 1883, at an international conference in Washington, the system of time zones was extended to encompass the entire earth.

The history of time measurement does not, of course, end with the standardization of time. With the world of space flights and nuclear fission comes the need to measure time with greater precision—and to fill that need, more accurate instruments and more minute divisions of time are required than were dreamt of only a while ago, as time goes.

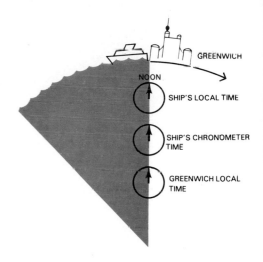

NAVIGATING BY THE CLOCK, a captain whose ship is moored near Greenwich, England *(above)*, synchronizes his chronometer with local time: Both the clock ashore and the sun's position show it is noon. After sailing westward *(below)* the captain finds that it is noon by the sun but 3 p.m. by his chronometer (i.e., Greenwich time). By allowing 15° for each hour's difference between his local time and Greenwich local time, he can fix his position: 45° west of his starting point.

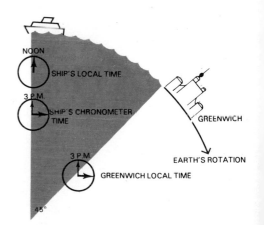

The Long Search for Perfect Timing

An ordinary watch ticks away the 604,800 seconds in a week with remarkable regularity: At the end of the week it may be off by only about 60 seconds. This amazing accuracy is the legacy of millennia of effort by philosophers, craftsmen and scientists. They sought a regular motion that could be counted to reckon the passage of time. At first the movement of the sun served. But because measurement of this natural motion is never convenient, and often impossible, man-made periodic motions had to be invented. The first known mechanical clocks, appearing in Europe in the 14th Century, achieved a measure of regularity from the oscillating motion of weights, which—governed by mechanisms called escapements—allowed the time indicators to advance periodically. But these instruments were crude and imprecise. Not until the 17th Century were two devices with truly regular motions invented. One was the pendulum, which swings back and forth with almost exactly equal periods. The other was the balance spring—a delicate wire that coils and uncoils with great regularity. Refined over the years, the spring has made possible a wide variety of accurate timepieces, from the ship captain's chronometer to the gift for the young graduate.

A CLOCK FOR NAVIGATORS
An early English chronometer, one of five built by the great clockmaker John Harrison in the period 1735 to 1770, included *(clockwise from top)* dials to indicate seconds, hours, days and minutes. This clock, Harrison's first, completed a highly successful test voyage to Lisbon in 1736. Harrison's fourth lost only 15 seconds on a five-month voyage to Jamaica.

Keeping Time with Shadows and Water

The daily progression of the sun's shadow, cast by a stick in the early days and later by such huge objects as obelisks, gave the first crude indication of the passing hours. But this was only the beginning of the search for a regular periodic motion.

The sun's shadow cannot be controlled to measure small time intervals, so the ancients used another natural steady motion—the dripping of water—to perform this function. With the water clock, or clepsydra, durations of time were reckoned by measuring the amounts of water that dripped into or out of a vessel. Unlike the obelisk, the clepsydra was a handy instrument: It could be carried from place to place and used in cloudy weather or at night. But still it had its drawbacks. It was hard to regulate and it froze in cold weather.

AN EARLY WATER CLOCK
In this Greek clepsydra, literally "water thief," time flowed as water trickled from one jar into another. Inside the jars, numbered scales or floats gauged the water level and the regular changes in the level measured intervals of time.

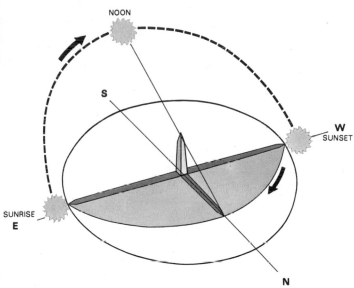

A 97-FOOT (29-M) TIMEPIECE

The obelisk at Karnak, Egypt *(left)*, built in about 1470 B.C., casts its shadow across the ruins of the temple of the sun god Amon Re. As the sun traveled from east to west *(broken line in sketch above)*, its shadow moved steadily in the opposite direction—from west to east—within the oval area around the obelisk. The three shadows shown here are those thrown by the obelisk at sunrise, noon and sunset.

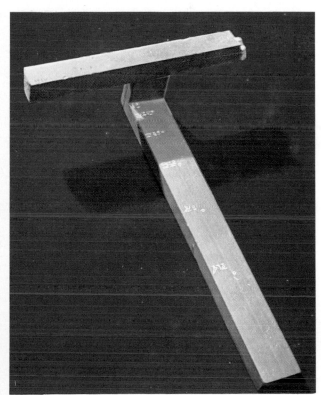

A SHADOW "WATCH"

A portable Egyptian shadow clock, built before the Eighth Century B.C., was pointed toward the sun so that the shadow of the crossbar fell on the hour scale on its handle. The scale included five hour lines plus the noon line. In the morning, the shadow clock was held with the crossbar toward the east; then it was turned to the west for the afternoon.

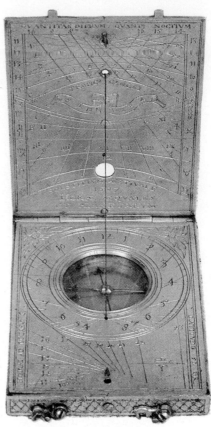

DIRECTIONAL TIME-TELLING

Compass sundials, introduced in the 15th Century, were the first shadow clocks to achieve both accuracy and portability. Several types were developed—including the folding sundial *(above)* and the plain cup *(below)*—but the method of operation of these timepieces was identical. The compass was used to point the sundial north, and the gnomon—a piece of string or a folding triangle—was lowered or raised to suit the latitude where the compass was being used. Both of the small dials shown here were designed to be carried in the pocket.

A PRECIOUS, MOVABLE SUNDIAL

This portable gold and silver sundial used in 10th Century England was held in the hand by its chain. The gold peg was inserted in one of the three holes on each side, each hole serving the two months listed below it (here March or October). The shadow of the peg was measured by the dots beneath, which marked four parts of the day, called tides in early England.

Enduring Timekeepers

As long as the principal timekeeper was the motion of the sun—for some 4,000 years from ancient Egypt to the 16th Century—the main timekeeping device was the sundial. It took innumerable shapes—flat dials, cube-shaped sundials, hollowed-out globes, flights of numbered steps (on which the shadow of a vertical wall was cast) and portable compass sundials. Yet the principle of all these timepieces remained the same: An object called a gnomon (Greek for "one that knows") casts a shadow on a numbered scale to show the hour.

For the shadow to move uniformly over the dial, the gnomon must be aligned parallel to the earth's axis. Since the earth's axis points in the direction of the North Star, and the North Star appears higher above the horizon in northern locations than in southern ones, the gnomon must be elevated to suit the latitude. At the equator, where the Pole or the North Star is low on the horizon, the gnomon has to be horizontal to point in the right direction; as the sundial is moved north, its gnomon has to be elevated; at the Pole, it is pointed straight up.

With the proper latitude adjustment, the sundial is remarkably reliable, so good that for centuries it was employed to correct mechanical clocks. Yet by modern standards the instrument is not truly precise; an ordinary sundial can be read only to within a minute of the precise time.

A DECORATIVE SUNDIAL
This iron sundial—an example of its most familiar form—was used in 18th Century Flanders gardens. It is decorated with idealized leaves and reclining figures in the ornate style of the time. Because the shadow moves more slowly over the dial at midday when the sun is overhead, the numbers on the face are placed closer together toward noon, farther apart in the early morning and in the late afternoon. The triangular gnomon is set for the latitude of Flanders, so the sundial, in its present location in New York City, could no longer show correct time.

Strange Ways of Reckoning Hours

As men stretched their imaginations to adapt regular motions of one sort or another to timekeeping, they invented many a strange technique. They measured time with the aid of burning candles, sifting sand, falling stones and running water. Occasionally they combined several mechanisms into elaborate Rube Goldberg contraptions.

One of the most remarkable of all these ingenious devices was a giant water clock built in 1357 A.D. in Fez, Morocco. The facade of this bizarre creation was 37 feet (11.3 m) long; the clockwork, which has since disappeared, is believed to have occupied a whole room and to have proclaimed the hour with a cacophonous clanging of gongs and creaking of doors.

Upon examination, some of the odd inventions of the ancients reveal elements that were to come together in more sophisticated form much later in the modern clock. A part of a Greek astronomical calculating machine proved to be an ancestor of today's clock mechanism. A huge Chinese water clock of the 11th Century A.D. included a device called an escapement, which later was to become the mechanical clock's first regulator.

A PRECURSOR OF THE MODERN CLOCK
Heavily encrusted after 2,000 years in the hold of a sunken ship, this fragment of a Grecian calculator was brought up from the Aegean Sea in 1900. When scientists attempted to reconstruct the calculator, which had once been used to compute the positions of the sun, moon and planets, they discovered that it had utilized gears like those used in later mechanical clocks.

THE PEBBLE CLOCK OF FEZ

Only the façade remains of a gargantuan 14th Century clock in Fez, Morocco, but from this remnant researchers have been able to reconstruct how the timekeeping machine worked. Inside the building a water clock, similar to a Greek clepsydra, activated a mechanism that released stone pellets from the wooden spouts below the roof line *(top)*. The pellets fell, one an hour, onto gongs 20 feet (6.1 m) below. If a passerby missed the chime, the same mechanism opened one of 12 doors *(in arches, center)* whose position indicated the hour. The door remained open until the next hour arrived.

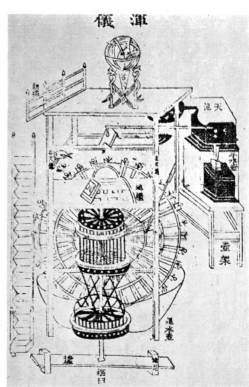

THE ESCAPEMENT'S ANCESTRY

In this Chinese water clock, designed in 1088 A.D. by Su Sung, water flowed from the spout of a reservoir at right to fill the buckets of the 11-foot (3.4-m) waterwheel. The escapement, a revolutionary novelty, engaged the large wheel to halt it until each bucket filled and the weight of the water turned the wheel another notch. Every quarter hour gongs clanged, trumpets sounded, and the wheel activated an elaborate series of carved figures which popped into view holding placards that told the time.

What Makes the Clock Tick

With the appearance of the first mechanical clocks in Europe in the 14th Century, the search for regularity moved closer to its goal. These instruments, which were designed more to display the movements of the planets than to show the hour, incorporated a crucial device, similar to that in Su Sung's water clock *(page 91)*: the verge and foliot escapement.

The foliot, a weighted bar, oscillated back and forth to bring a measure of regularity to the clock. Moving with it, the verge intermittently engaged teeth connected to the clock's driving mechanism, thereby stopping and releasing the clockwork so that it moved ahead tooth by tooth.

It was this action that put the tick in the clock. Yet, ingenious as it was, the verge and foliot escapement had no independent source of motion but had to be driven by the same weights that drove the rest of the clock. This factor plus friction and the crudeness of handmade parts made these early mechanical timekeepers irregular. Their descendants, the earliest true clocks, were equipped with one hand.

PROTOTYPE OF ALARM CLOCKS
A German wall clock of the 14th Century was a small version of the early mechanical clock. Regulated by a verge and foliot escapement *(weighted bar at top)*, it was driven by a weight that hung down below the clock. The wall clock was employed to toll the hours in monasteries.

HOW THE ESCAPEMENT WORKS
The verge and foliot escapement *(blue)* both regulated the drive mechanism *(red)* and was actuated by it. As the weighted foliot bar at the top oscillated back and forth, the vertical rod, or verge, below it turned and engaged first one, then the other, of its two pallets, with the teeth of the crown wheel. At the left, the top pallet has engaged the wheel, momentarily halting it. Then as the verge *(center)* and foliot bar are turned again by the crown wheel, the top pallet swings free. At right, the bottom pallet stops the crown wheel before starting to swing back.

THE OLDEST MECHANICAL TIMEKEEPER.
Built in 1386 for the cathedral tower at Salisbury, England, this massive instrument is probably the oldest mechanical clock in existence. At center is its saw-toothed crown wheel, geared to the clock hand and to the weighted rope drive but regulated by the verge (vertical bar) and foliot (horizontal rod on top of the verge).

An exposed view of Jacob Zech's clock shows the cone-shaped fusee at upper left; the drive spring is inside the engraved drum *(upper right)*.

A Spring for Portability

In the early 16th Century an ingenious locksmith in Nuremberg, Germany, named Peter Henlein built one of the first clocks driven by a wound iron spring instead of by the larger, clumsy weights previously used in mechanical clocks. Henlein's small, egg-shaped clock was the first watch.

Henlein's watches—later called "Nuremberg eggs"—became immensely popular with Europe's wealthy classes. Fashionable people had them made of gold and silver and set with precious stones. But the "eggs" kept time poorly. The spring's driving force changed as it unwound; as a result, the timepiece ran too fast when the spring was taut and then gradually ran down like a child's spring-driven toy. The first known instrument to overcome this crucial weakness was designed by a Czech named Jacob Zech. His spring-driven clock included the fusee (below, right), a cone-shaped device that evened out the force reaching the clockwork. This improvement converted the mainspring into a practical power source, but the watch still was regulated by the old unreliable verge and foliot escapement. The quest for regularity, which had been concerned up to this time only with the clock, turned now to portable timepieces as well.

A ONE-ARMED TIMEKEEPER
The face of Zech's clock supplied a variety of information even though the clock had only one hand. The outer circle numbers the hours I through XII twice, while the dial next to it re-numbers the hours from 1 to 24. Moving closer to the center are the 12 signs of the zodiac. The small indicator at the top keeps track of the "age" of the moon from 1 to 29½ days.

THE FUSEE'S CLEVER SHAPE
The driving mechanism of Zech's clock included a cone-shaped fusee and a spring (inside drum at right). As the spring uncoiled, it turned the fusee and the watch's gears by a cord. The cone shape of the fusee compensated for the changes in the spring's force. When the spring was coiled, it pulled vigorously on the tapered top of the cone, where the leverage was poor. But as the spring unwound and its pull weakened, the increased diameter of the lower part of the fusee gave greater leverage to maintain a fairly uniform turning motion of the clockwork.

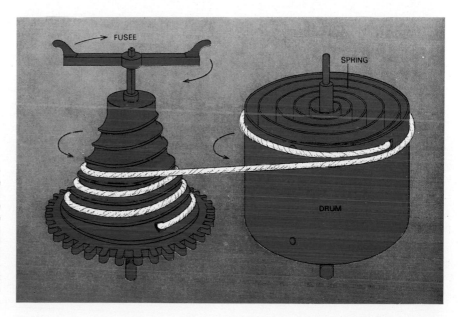

EARLY SPRING-DRIVEN PENDULUM CLOCK

HUYGENS' 1673 CLOCK

Huygens' pendulum clock design was weight-driven *(below left, red)* through a crown wheel engaged by a verge *(blue)*. The verge was turned by its pendulum connection—an L-shaped rod having an eye for the pendulum bar *(detail at right, below)*. To make the bar swing in a steeper curve than could happen naturally, Huygens added bent stops to deflect the suspension string.

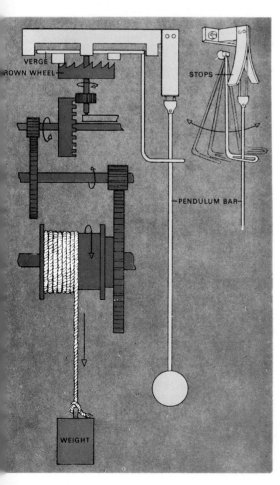

VERGE

CROWN WHEEL

STOPS

PENDULUM BAR

WEIGHT

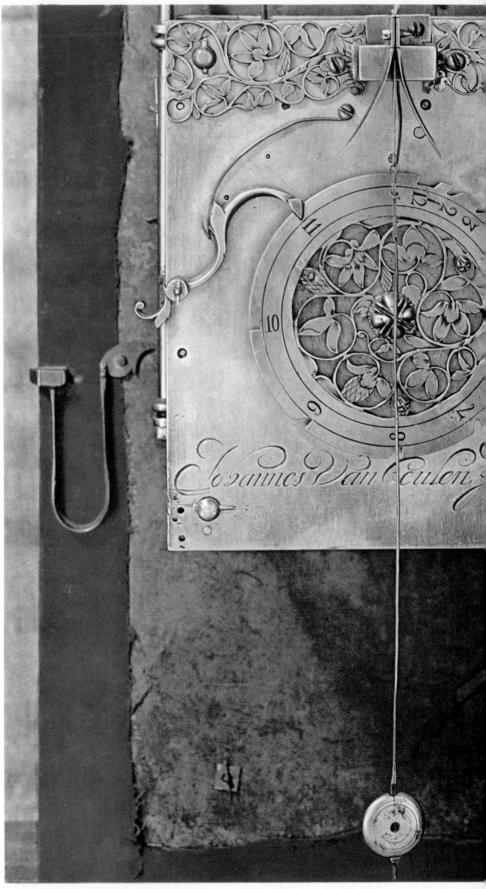

The pendulum and chime-controlling dial show in a rear view of the clock designed by Huygens,

The Pendulum's Regular Rhythm

Early in the 17th Century, Galileo conceived an idea that finally was to bring regularity to the clock. As his biographer, Vincenzo Viviani, told the story, Galileo observed a lamp swinging from the ceiling in the cathedral of Pisa and noticed that the lamp always seemed to take the same time to swing back and forth. Being a medical student, Galileo timed the lamp with his pulse and decided that the pendulum's nearly perfect period would provide an ideal regulator for a clock. The story makes good reading and it has become embedded in the lore of the clock's development, but it is mostly fiction, a creation of Viviani who was Galileo's student and admirer, as well as his biographer. Actually, Galileo first became interested in the pendulum in the course of some studies of the effect of gravity. When the idea of adapting a pendulum to a clock occurred to him, he was an old man and nearly blind. He made some sketches and an incomplete model, but it was the Dutch genius Christian Huygens who built the first clock regulated by a pendulum. It brought greatly increased accuracy partly because, unlike the foliot bar, it was not moved primarily by the clock's driving mechanism. The pendulum was chiefly actuated by the force of gravity. As a result, Huygens' clocks with their free-swinging pendulums were the first timekeepers able to count seconds.

Before the pendulum clock could be perfected, many refinements were needed. Temperature had a critical effect. Rising temperature expands the metal in the pendulum rod, causing it to swing more slowly; with a 39.37-inch (100-cm) steel rod a rise of 10° (5.6° C.) causes a loss of two and a half seconds a day. Various materials were tried to compensate for this expansion; today a nickel-steel alloy, which does not expand perceptibly, is used. With this and other improvements, such as the anchor escapement, pendulum clocks achieved so high a degree of regularity by the 18th Century that they were keeping time to within a few seconds a week.

THE PENDULUM'S NEW ESCAPEMENT
The anchor escapement *(blue in diagram)* is regulated by the pendulum. In sequence, the left-hand pallet engages the crown wheel *(top, red)*, then the pendulum swings this pallet free, the crown wheel advances, and the right-hand pallet in turn comes down. Because the anchor touched the crown wheel more delicately than the verge, it interfered less with the pendulum.

and built by his associate Johannes van Ceulen.

97

The Spring That Times a Watch

Within 20 years of the time when he built the first pendulum clock, Christian Huygens also designed a nearly perfect regulator for the watch. Called the balance spring, this precise mechanism uncoils and coils to move the balance wheel back and forth and thereby turn the pallets. The key to the balance spring is the regularity with which it frees the pallets from the crown wheel, a regularity comparable to that of the pendulum. With the discovery of the pendulum and the balance spring, the quest for regularity came to a climax. The search for even greater precision would go on: Electric clocks and atomic clocks that could reckon time in tiny fractions of a second were yet to appear, but a workaday regularity for clocks and watches had arrived.

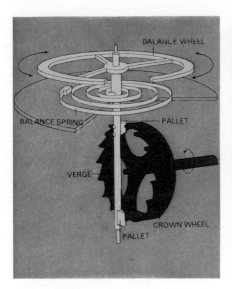

REGULATOR OF THE WATCH
The balance spring regulator is shown in action in this diagram. Like the pendulum, it is almost independent of the drive mechanism; as the balance spring *(blue coil)* unwinds and winds, it turns the balance wheel *(top)* back and forth with near-perfect regularity. As the balance turns, so do the verge and its pallets, which alternately engage the toothed crown wheel *(red)*.

TINY BUT TRUE TIMEPIECE
The precision and compactness of the balance spring made possible pocket watches such as the one above designed by the English clockmaker, Daniel Le Count, in 1676. Only two and a quarter inches (5.7 cm) in diameter, the watch was accurate enough to include a dial in minutes *(right)*.

5
Segments of the Second

Fracturing time in art, Marcel Duchamp's 1912 painting, *Nude Descending a Staircase*, anticipated a scientific technique perfected 20 years later: the time-analyzing strobe photo *(page 109)*, which portrays a moving figure at several different instants.

ON A HILLSIDE in Washington, D.C., stands a small, square, windowless shed. Inside it is a very peculiar sort of telescope—one forever fixed in a vertical position. Nobody looks at the stars with this telescope, or tracks satellites, or maps the face of the moon. It was designed and built for one task only: to determine the moment at which certain stars cross the zenith—the point in the heavens directly overhead.

This curious instrument is the Photographic Zenith Tube—informally, "the PZT"—of the United States Naval Observatory. The PZT's photographs of star positions clock the rotation of the earth and provide the absolute criterion of Standard Time in the U.S. When anyone in this country, or its possessions, or its naval vessels, or its military aircraft, asks, "What time is it?" the answer ultimately derives from the PZT. With its aid, Naval Observatory scientists keep their own master clock (and the broadcast time signals it controls) in time with the spinning earth to within eight tenths of a second. The master clock, operated by the vibrations of cesium atoms, divides the 24 hours between PZT sightings into intervals that are accurate to several billionths of a second a day.

It is hard for most people to imagine why anyone would *want* to measure time in billionths of a second (nanoseconds). In our daily affairs—taking a cake out of the oven, getting to school, catching the morning train—our clocks and watches serve us well if they are correct to within a minute or two. Even so specialized an activity as timing a footrace requires a stopwatch accurate only to tenths of a second.

Yet almost everyone relies, directly or indirectly, on ultraprecise time measurements. They are essential to the distribution of electric power; the utility engineer, synchronizing the output of tens or hundreds of generators, needs a master clock correct to a few tens of microseconds (millionths of a second). In transportation—particularly at sea—accurate timekeeping can be a matter of life or death; the ship's captain groping through a fog may be depending on radio signals synchronized to better than a microsecond. Without precise time measurement, modern science could not exist. The physicist studying the pieces of atoms produced by an atom smasher deals in particles that appear and disappear in nanoseconds or picoseconds (1/1,000 of a nanosecond); to make sense out of his experiments, he must be able to determine the time sequence of these fleeting events with split-microsecond precision.

The timekeeping needs of modern civilization have stimulated the development of clocks that keep time more accurately than the earth itself, operating with a gain or loss of no more than one second in 350,000 years. Time, indeed, can be measured at least a thousand times more accurately than length, temperature or any other physical quantity, and the experts hope before long to measure it many times more accurately still.

What has made precise time measurement essential is primarily the use of electricity and electromagnetic radiation. Electricity powers much of our civilization; electricity piped through telephone and telegraph wires, electromagnetic waves broadcast through the air, provide us with near instantaneous communication. Other types of electromagnetic waves—the microwaves of radar and the far smaller waves of X-rays—permit us to discern otherwise invisible features of the world we live in.

In dealing with electricity and electromagnetic radiation, precise time measurements are important for two reasons. First, both electricity and electromagnetic waves move fast. In the seemingly trivial period of one microsecond, a radio wave or a radar pulse can cover 1,000 feet (305 m).

But the second and more basic reason for precise time measurements lies in the nature of electrical phenomena. They involve electrical and magnetic forces which shift back and forth very rapidly, tens or thousands or billions of times a second. The "frequency" of this shift—measured in cycles, kilo (thousand) cycles or mega (million) cycles per second—defines the types of electricity and electromagnetism. Thus ordinary household alternating current is rated at 60 cycles per second, long-wave broadcast waves are measured in kilocycles, FM and TV waves in megacycles. Radar takes us into the range of thousands of megacycles, X-rays into still-higher frequencies.

Because frequency is so basic, its control becomes a matter of great practical necessity. Rather small errors in frequency can black out a power system, garble a Teletype message or blur a radio signal. But the precise standardization and control of cycles or megacycles per second imply equally precise measurements of seconds—and microseconds.

The electric power industry provides a good example of the importance of frequency control. A modern power network may include hundreds of generators, all of which must operate harmoniously. If they do not, the generators tug against one another. One engineer has compared a power system to a group of heavy pendulums linked together by fragile threads: Unless the pendulums swing in unison, the connections will break.

Bad timing makes the lights go out

To swing in unison, the generators of a power system, like the pendulums, must start swinging together. But more important, they must swing with the same rhythm—operate at the same frequency—or they will not long continue together. Minor frequency variations in a system tend to correct themselves automatically; one generator "pulls" another back into step. But a drastic frequency shift, resulting from a sudden and massive drop in the availability of power at some point, can undermine the stability of the whole system. Great surges of power jump from point to point, burning out equipment; under certain conditions, the whole system may shut down. Residents of the Northeastern states discovered this to their sorrow on November 9, 1965, "the night of the big blackout," when six states were plunged into darkness for as long as 11 hours.

To keep an entire system in step, power companies continually check frequencies against electronic devices that generate very accurate standard frequencies; these are, in effect, master clocks. Most companies try to maintain a "60-cycle" output between 59.98 and 60.02—that is, with an error of no more than one part in 3,000. The master clock must be a great deal more accurate—its errors are held to one part in 10,000.

SCIENTISTS' TIME MEASURES—frequency and phase—associated most often with electrical phenomena are illustrated above by the marching soldiers. In all three columns, the squads pass the reviewing stand one after the other and at identical intervals. In electrical terms, all three columns are marching with the same frequency, just as electric currents whose alternations are equally spaced in time are said to share the same frequency. Only the upper two columns are synchronized. Their squads pass the reviewing stand at the same instant, and are said to be in phase, while the squads of the lower column pass the stand at a different moment and are out of phase with the others. Similarly, the alternations of electricity are in phase if they occur at the same instant but out of phase if they do not.

Though power companies insist that they are selling *power*—not accurate frequency or accurate time—the precision of their frequency control does in fact determine the accuracy of any electric clock running off the power lines. A frequency shift up or down means a slight speeding or slowing of the clock. These slight differences in speed do not always cancel out completely; over several hours they may add up so that the clock may be as much as two or three seconds fast or slow.

Most clock owners are not aware of such slight changes. But for some enterprises, such as broadcasting networks, whose programs are synchronized almost to the second, even the smallest change is significant. For years broadcasters used ordinary electric clocks in their studios and control rooms, until some embarrassing experiences forced them to employ more reliable time standards.

During World War II, when U.S. power supplies were strained by heavy demands from defense industries, large errors in power frequencies occurred. Moreover, the greatest frequency shift happened at different times in different time zones. As a result, clocks in, say, New York and Los Angeles might differ from one another by as much as 20 seconds.

Keeping time with a tuning fork

If the Los Angeles studio clocks were ahead of New York, an announcement from California that "we now take you to our New York studios" would be followed by a 20-second silence while New York, watching its laggard timepiece, waited to begin its program. If Los Angeles was behind New York, the result was worse, since the switchover would take place only after the New York program was already underway. As a result, Pacific Coast listeners might miss an entire commercial.

Broadcasting engineers solved this problem by means of a device previously confined to the musical world—the tuning fork. When a tuning fork is struck, it sounds a steady musical note because its arms vibrate at a remarkably steady frequency. The vibrating arms, the engineers saw, could be used as the moving mechanical parts of a device that operates almost like an electric generator, producing a current at the exact frequency characteristic of the fork. By building a fork that would vibrate at 60 cycles per second, the engineers obtained very precise 60-cycle power for clocks in control rooms across the country, causing them to run on the same unchanging time. This method kept the network "in time with itself" to within a second or better.

Even more basic to the broadcasting business than accurate time in the control room is accurate frequency of the broadcast signal. If the frequency "drifts"—as it did during the early days of broadcasting—listeners must retune their sets periodically. More serious is the confusion caused when a drifting station edges into another station's channel. To prevent such interference, broadcasters are required to maintain their assigned frequencies within narrow limits: WNBC, New York, kingpin of NBC's radio operations, broadcasts on 660 kilocycles with a permissible error of no more than 20 *cycles*—one part in 33,000.

To control their frequencies, broadcasters have long used a mechanism that vibrates at frequencies too high for humans to hear. This is the quartz crystal oscillator. A properly prepared quartz crystal can be made to vibrate at a constant frequency, one that is stable to within a few parts per billion. While vibrating mechanically, such a crystal also generates a weak electric current of the same constant frequency, and this current can be used to set the frequency of a broadcast wave.

A clock that loses a millionth of a second a day

The stable frequency current generated by a quartz crystal, like that induced by a tuning fork, can also be made to power an electric clock. The first crystal clock, built in 1928 by W.A. Marrison of Bell Telephone Laboratories, immediately achieved an accuracy of 1/1,000 of a second a day—nearly 10 times as good as the best pendulum clocks. By the 1970s the technology of quartz crystal clocks was so far advanced that wristwatches utilizing the same mechanism were being sold widely. In a casing no larger than that of a traditional wristwatch a tiny battery makes a crystal of quartz oscillate precisely 32,768 times a second. Then a system of electric circuits counts the quartz oscillations. When each second passes, an electric pulse causes an arrangement of small lights on the face of the watch to display the new time. These very simple quartz timekeepers gained or lost less than a sixth of a second a day. Far more elaborate crystal clocks, widely used for astronomical and other scientific purposes, are tuned until they are precise to a few microseconds a day.

When crystal clocks made possible the splitting of a second into a million parts, scientists once again asked themselves a question that had been propounded much earlier: "How long is a second?" The official definition was still the same one that a committee of French scientists had proposed in 1820—1/86,400 of a mean solar day. Thus the second was based on the same astronomical "clock" that mankind had been using since the Sumerian and Egyptian priests first numbered the hours: the earth's daily rotation on its axis.

But time, as even the Sumerians and the Egyptians knew, can also be measured by other astronomical clocks, notably the orbital rotation of the moon around the earth and of the earth around the sun. The orbital motions of Venus and Mercury serve equally well; so do the rotations of Jupiter's moons around their parent planet. All these celestial clocks operate independently of one another and of the earth's rotation.

In 1939, after a long and careful series of celestial observations, the English astronomer H. Spencer Jones startled scientists by announcing that the spinning earth was not keeping accurate time. Orbital time (technically known as ephemeris time), based on the earth's revolutions around the sun, did not vary, Jones pointed out. However, rotational

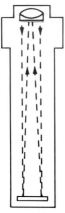

THE PHOTOGRAPHIC ZENITH TUBE (PZT) keeps track of the earth's rotational time by focusing light rays *(dotted lines)* from celestial objects passing overhead and photographing them. The master clock *(below)* times the photographs with split-second accuracy.

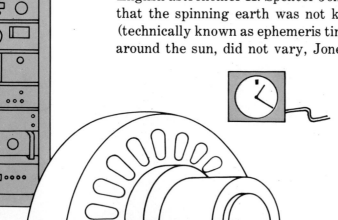

A CHAIN OF TIMEKEEPING DEVICES is synchronized with the time recorded by the master clock. Power generators *(left center)* emit alternating current at precisely 60 cycles per second. This frequency controls the standard time displayed by household electric clocks *(left)*. Standard Time is derived from the master clock's Mean Solar Time by adding the number of hours appropriate for a particular time zone. But the master clock remains an accurate index of Mean Solar Time only if it is periodically readjusted according to the readings of the PZT.

time, based on the spin of the earth upon its axis, varied noticeably.

Many kinds of observations have confirmed that the earth does indeed spin a little faster at some times than at others. One cause is the wind. The pattern of prevailing winds over the earth changes from season to season and so do wind velocities, which average higher in winter than in summer. And wind, through friction with the earth's surface, can accelerate or retard the earth's spin very slightly.

Water is also thought to play a role. In winter, millions of tons of water are temporarily removed from the oceans and deposited, as snow and ice, on polar and temperate lands. In the process, the water is lifted up hundreds or even thousands of feet above sea level. The transfer of this mass outward slows the earth's rotation slightly, just as a twirling ice skater can slow his spin by moving his arms outward.

How the tide slows the earth's spin

These changes are small and they tend to balance themselves from season to season. But they are big enough to mask what may be a far smaller change in the earth's rotation, yet one that is constant and therefore cumulative over the long pull: the slowing of the earth by tidal friction. Geophysicists believe that tidal currents in the oceans, hitting the sea bottom in shallow water, retard the earth's rotation slightly.

Scientists have found plausible evidence for the cumulative effect of tidal slowing in many different phenomena. A certain type of marine animal that flourished about 350 million years ago produced layered deposits on its shell, laid down according to a rhythm governed by the amounts of both sunlight and moonlight it received. Fossils of these animals with the shell layers still intact indicate that the earth was rotating much faster then: The length of the day seems to have been about 20 hours instead of 24, and the month was only 28 or 29 days long. Other evidence for the earth's slowing comes from ancient records of eclipses. An eclipse is visible over only a small part of the earth's surface. Moreover, the area of visibility can be calculated for eclipses that occurred centuries (or even millennia) in the past. It turns out, however, that modern calculations do not agree with the ancient records. The eclipses seem to have been observed in areas some hundreds of miles to the east of where they should have appeared.

Since eclipses depend on orbital motions of the sun, moon and earth, they "run" on orbital time. But their areas of visibility, which depend on what part of the rotating earth's surface is in the right position for seeing the eclipse, are governed by rotational time. Thus if the area of visibility was not where it "should" have been, it follows that rotational time was out of step with orbital time.

Such variations in rotational time ruled out the mean solar day as a basis for defining the second. In 1956 the international definition of the second was changed: A second was redefined as 1/31,556,925.9747 of the orbital year that began at noon on January 1, 1900. As far as scientists could tell, this second was an invariant quantity, but very accurate measurements took so long to make that some uncertainty still remained. Fortunately, as some scientists agonized over the length of the orbital year, others were perfecting their acquaintance with another, far more convenient and impeccably precise tool for measuring time—the atom.

Like pendulums, individual atoms and the ordered groupings of atoms called molecules can serve as standards of frequency—and therefore of time. When a pendulum is pushed it begins to oscillate; similarly, when an atom or molecule is exposed to X-rays, visible light, or other kinds of electromagnetic radiation its components begin to move in unaccustomed paths. An electron orbiting inside an atom may jump from one position to another, it may change direction or it may wobble on its axis like a top. All the atoms inside a molecule may jiggle like balls attached to springs. These atomic movements are of great interest because they are produced by very specific frequencies of electromagnetic radiation, called resonance frequencies. Each atom will absorb radiation of certain resonance frequencies only.

Scientists therefore realized that they could prepare a quartz crystal to generate a beam of electromagnetic radiation and use a specific atom or molecule to insure that its frequency never varied. The first clock based on an atomic frequency standard was produced by scientists at the National Bureau of Standards in 1949. In it a quartz crystal was kept tuned by a chamber containing a small number of ammonia molecules. A similar clock using cesium atoms instead of ammonia molecules was built in 1952, and successive improvements were made until in 1969 a cesium clock was devised which was accurate to a few billionths of a second a day, a new record of timekeeping precision.

Harnessing the cesium atom to a clock

Cesium, a soft, silvery-white metal, was selected for the atomic clock because one of its electrons is particularly vulnerable to electromagnetic radiation. If the atom is bombarded with microwaves moving at precisely 9,192,631,770 cycles per second, this electron will begin to spin much differently from the way it usually does. The entire energized atom will then respond differently to a magnetic field than a regular cesium atom does. In the cesium clock, a vibrating quartz crystal is made to generate microwaves at the cesium resonance frequency. These waves are passed through a chamber containing cesium atoms, which then leave the chamber through a magnetic field. If the microwaves are precisely at resonance frequency, the beam of cesium atoms will be bent by the magnetic field and will strike an electrical detector. If the microwaves are slightly off resonance frequency, the cesium atoms will not absorb the radiation, and the beam will miss the detector. An electrical signal will then be sent to the quartz, indicating that it is vibrating at the wrong frequency, and a circuit will be activated to speed it up or slow it down. The regulated quartz oscillations form the frequency standard for an electronic clock, which counts them and displays the result in multiples of a second.

Variations on the cesium clock using other atomic resonance frequencies have been tried. One of them, a quartz-crystal clock regulated in a similar way by hydrogen atoms, and called the hydrogen maser, actually proved to be about 100 times more precise than the cesium clock for periods of time shorter than a day. After running for more than a day, though, the maser's stability decreased, leaving to the cesium clock the title of the most consistently stable timepiece ever invented. Cesium clocks are the basis of the precise time and frequency standards of both the Naval Observatory and the National Bureau of Standards.

The cesium clock had such an impressive record that in 1967 an international committee changed the definition of the second once more. For the first time in history the second was made independent of movements of the sun and the earth. It was defined as 9,192,631,770 natural periods of the cesium atom, or the time in which that many cycles of electromagnetic radiation must strike a cesium atom in order for it to be catapulted into its higher energy state.

Timing the passage of the stars

This new second is precise enough for scientists and engineers measuring very short-lived phenomena. However, the ship or plane captain determining his position and the astronomer charting the orbit of a satellite have different needs. They must know the time in relation to the earth's longitude, in order to answer the question: "How long ago was the celestial object, now overhead at my location, overhead at the point of zero longitude (Greenwich, England)?" They need to measure not simply time intervals but time intervals related to the earth's rotation.

It is here that the Naval Observatory's PZT, with which this discussion began, comes in. Every night, weather permitting, a number of stars are photographed through the tube as they pass overhead; the photographs are timed through a link with the master clock. Calculations tell the observatory staff where the stars "should" have been at the time they were photographed; the photograph tells where they actually were. The differences in position, measured under a microscope, give an accurate index of how far the earth's rotation has wandered from the master clock's atomic standard. The differences from night to night rarely amount to more than a few thousandths of a second, but when the total error nears 8/10 of a second, the master clock must be realigned. Therefore, about once a year, the last hour of the last day of a month (usually December) has an extra second appended to it, as the mechanisms of the master clock and the time signals it controls are stopped for precisely one second. These "leap seconds," when time stands still, serve to keep Mean Solar time— the earth's rotational time—in phase with the dazzlingly precise man-made time of the master clock.

Measuring What
the Eye Cannot See

Like the instant when a rocket's fuel ignites to hurl a spacecraft into orbit, many of the intriguing events of modern life happen too quickly for the eye to see. No one can tell by looking at a lightning flash, for example, that the main bolt travels upward, from the ground to the clouds, and not downward. No eye is fast enough to register a clear image of a spinning airplane propeller, a bullet arcing from a rifle muzzle or an atom splitting into the energized particles that drive nuclear submarines and power plants.

In past centuries, the very conception of measuring brief intervals was lacking. It was not until the 18th Century that second hands regularly appeared on the faces of clocks. But the ever-increasing speeds of an industrial age and a mounting curiosity about the fast movements of nature have impelled modern scientists to investigate smaller and smaller segments of time. The techniques they have devised for slicing seconds are now so refined that it is possible to follow step by step the progress of the explosion of a stick of dynamite, to count the rapid beating of a hummingbird's wings — even to chronicle the lives of fleeting subatomic particles that appear and then disappear within billionths of a second.

TIME AS MOTION
This multiple-exposure photograph times a drum majorette's whirling baton. A brilliant light, flashing once every sixtieth of a second, has caught the baton in successive positions as it tumbled through the air. A count of the baton's images—which coincide with the light flashes—reveals the time it took in spinning through one complete turn: exactly .4 second.

TENTHS OF A SECOND
The Exactly Timed Photo Finish

In the world of sports, split-second intervals of time often mean the difference between obscurity and a spot in the record books. In a track meet races are clocked officially to a tenth of a second; frequently, however, a still smaller interval separates the winner from the runner-up.

In judging a close race, track offi-

A photo-finish record shows winner Dwight Middleton and five also-rans crossing the tape in the 440-yard (402-m) dash at the 1966 U.S.

cials depend on timed photographs, which are far more accurate than the eye in determining who crossed the finish line first. Though the photographs may look like pictures taken by a conventional camera, they are quite different. Instead of the one moment of time normally caught by a snapping shutter, they show a succession of moments as contestants cross the tape, one after another.

A photo-finish camera—such as the Bulova Phototimer used to take the picture below—records time as distance, right to left, across the film. This type of camera has no shutter but is focused along the finish line through a thin vertical slit. A strip of film rolled behind the slit at a constant rate of speed records a continuous image of the finish line, catching each runner in turn as he crosses the line. Inside the Bulova camera a counter clicks off the hundredths of seconds. The camera photographs the counter, recording the runner's time above his picture on the film.

Track and Field Federation Championships. The numbers at the top of the photo give Middleton's exact unofficial time: 46.42 seconds.

Clocking a Nerve Impulse

To spot diseases which attack the nervous system, doctors time the velocity of nerve impulses. Normally in the larger nerves impulses travel a yard (0.9 m) in a few hundredths of a second. A reduction in this rate indicates damage—which might result from such disorders as diabetes, lead poisoning or extreme alcoholism.

The nerve signal is electrical, a weak jolt which travels along the body's nerve cables. To clock its pro-gress, doctors trigger an impulse arti-ficially by stimulating the nerve with electricity. They measure the time it takes the impulses to travel from one point on the nerve to another. Their clock is an oscilloscope—an electronic device which translates time intervals into patterns on a screen *(right)*.

One of the nerves doctors test is the ulnar nerve—the sensitive one that tingles when the "funny bone" is hit. Since the distance from spinal cord to hand along this nerve is about three feet, an impulse should take about 0.02 second to reach the hand, where the signal activates the mus-cles which move the little finger. But even mild damage from disease can double the transmission time. The amount of retardation tells a doctor how badly the nerve is damaged. Then, when the nerve is healing, he gauges its recovery by noting the im-provement in the transmission time.

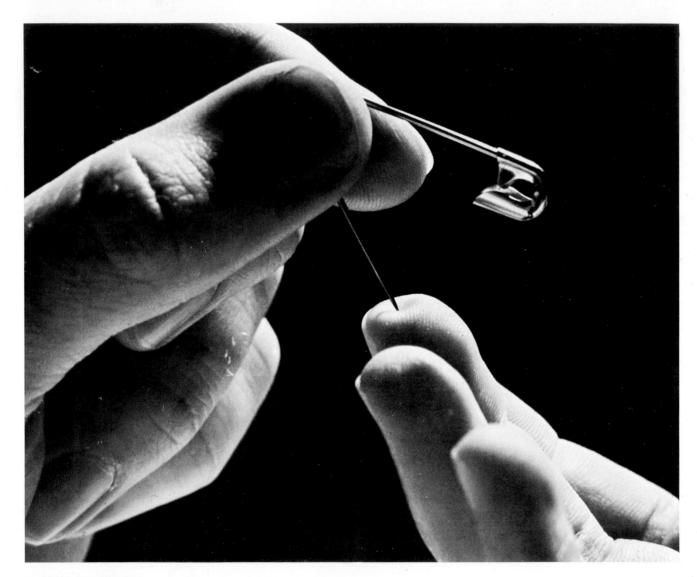

TRIGGERING A REFLEX
Stabbed by the point of a pin, this hand will jump back in an automatic reflex even before the first twinge of pain reaches the brain. The discrepan-cy in timing results from the different speeds with which various sensory nerves transmit sig-nals to the spinal cord and the brain. The impulses which cause the reflex move along broad nerve channels, traveling to the spinal cord and back in hundredths of a second. But those which cause the sensation of pain travel more slowly; it may take half a second before they register in the conscious areas of the brain.

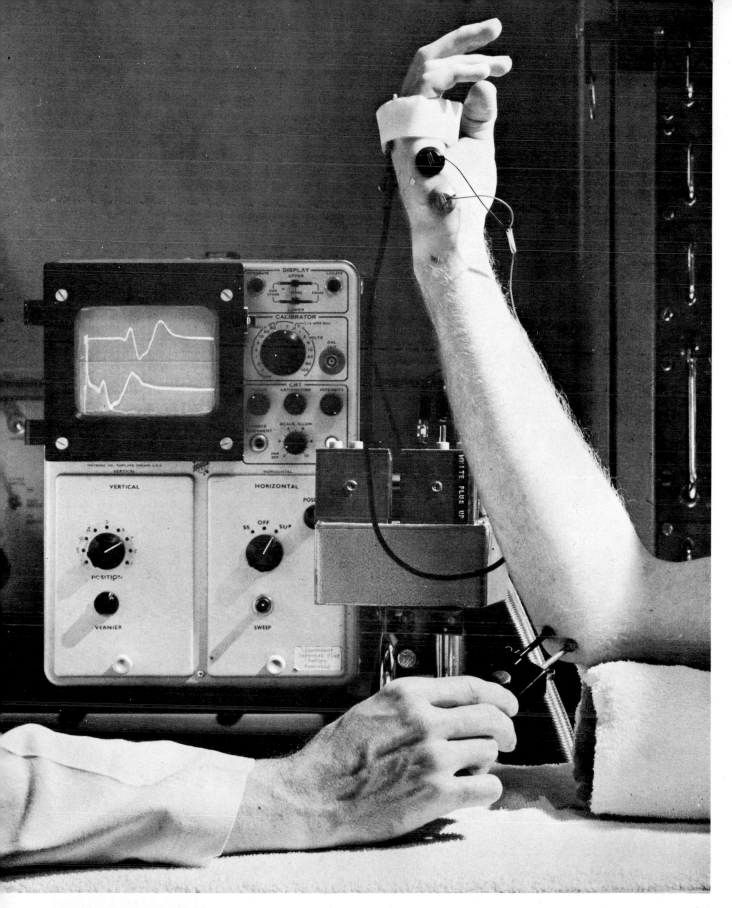

GRAPHS THAT CLOCK A NERVE

Graphs on the screen of an oscilloscope *(upper left)* trace the speed of impulses through the ulnar nerve. In this test doctors need to measure the speed of an impulse from the elbow to the wrist. Since the impulse alone is too weak to detect, they record the twitching of a hand muscle, which is indicated by the peaks and valleys on the screen. This method requires two measurements: one from elbow to hand muscle *(top trace)*, and another from wrist to hand muscle *(bottom trace)*. Simple subtraction then gives the time an impulse takes from elbow to wrist

The flash of a high-speed strobe light reveals three successive stages in the bursting of these three balloons as they are ripped into shreds

THOUSANDTHS OF A SECOND
A Light Flash to Transfix Time

Bullets whiz through the air at more than 1,000 feet (305 m) per second, arcing and spinning in ways that affect their range and accuracy. Raindrops erode the earth by almost instantaneous splattering. The moment of impact between the head of a golf club and a golf ball lasts only a two thousandth of a second; what

in two thousandths of a second by a 22-caliber rifle bullet—itself caught in flight at far right as it travels at 1,200 feet (366 m) a second.

happens in that moment determines the numbers on the final scorecard. None of these occurrences can be seen by the unaided eye.

Yet it is possible to stop and record such events—which are important to engineers, geologists and manufacturers of sporting goods—with the help of lamps that flash intensely but very briefly. These electronic flash guns are familiar equipment for photographers. A sophisticated version of the electronic flash is the stroboscope, used in industry to detect flaws in fast-moving machinery. The strobe, as it is often called, emits its short bursts of light at regular, rapid intervals. Some models can flash more than 100,000 times a minute. If the tempo of the flashes is set to coincide with the rotations of a machine part—say a spinning fan blade or airplane propeller—the moving part is always in the same position when illuminated and it looks as though it were standing still, like spokes of a turning wagon wheel in a Western movie.

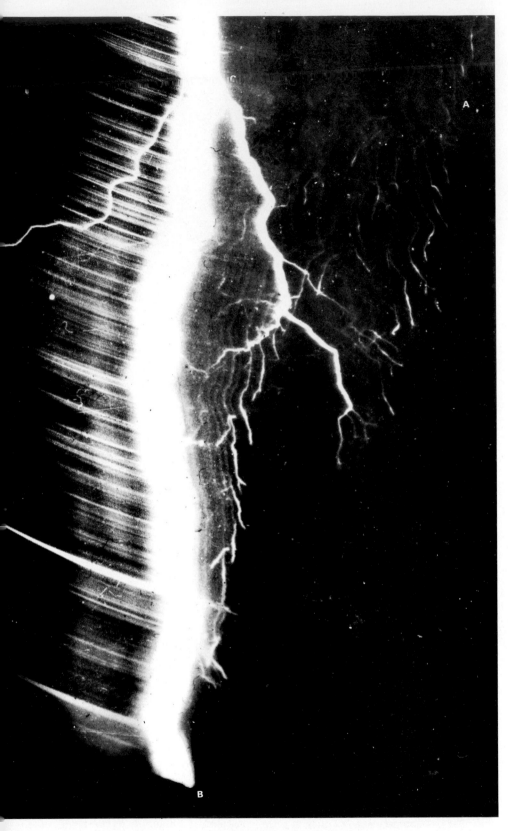

A

B

TEN-THOUSANDTHS OF A SECOND

A Camera That Dissects Lightning

Almost two centuries after Benjamin Franklin flew a kite in a thunderstorm and drew "electrical fire from clouds," no one really knew what happened when lightning slashed the sky. Not until the invention of a special camera, able to break down the flash of electricity into a meaningful time sequence, could scientists dissect the anatomy of a lightning stroke.

The camera, which uses a moving lens to spread the image of a lightning bolt across the film *(left)*, was devised by Sir Charles Vernon Boys in 1902. Photographs taken with Boys's camera showed that the main bolt is preceded by a "leader"—a dim preliminary stroke that forks its way down from the cloud to the ground to etch a path for the main discharges. The leader may take a hundredth of a second to reach the ground, but the main discharges, a series of brilliant strokes traveling from the ground up to the thundercloud, each take less than a ten thousandth of a second.

A BOLT FROM THE GROUND UP
This conventional photograph of a lightning flash *(right)* shows only the brilliant main discharge, which obliterates the dim image of the leader. The forks show where the first upward, return stroke has backtracked down blind alleys taken by the leader in its path to the earth.

A WELL-TIMED FLASH
The moving lens of a Boys camera has swept from right to left across a film to produce this picture of a lightning flash. By spreading out the image of the flash, the camera separates its components. The filaments to the right show the leader as it gropes its way in quick, hesitant darts toward the ground from point A to point B at a rate of 290 miles (467 km) a second. The streak at the left is the initial return stroke. That flash, the main visible stroke, traveled from B to C—a distance of some 2,000 feet (610 m)—at a speed of 37,500 miles (60,349 km) per second.

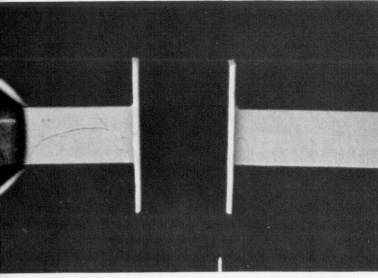

1. At 20,000 feet (6,096 m) a second a blast eats into a pentolite stick.

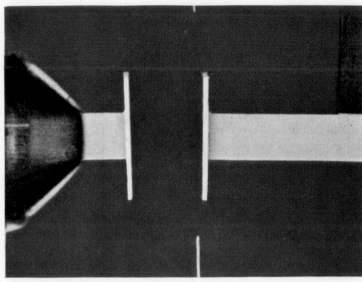

2. Two brilliant white shock waves angle out at the vanguard of the blast.

5. In three microseconds the explosion bridges the one-inch (2.5-cm) gap.

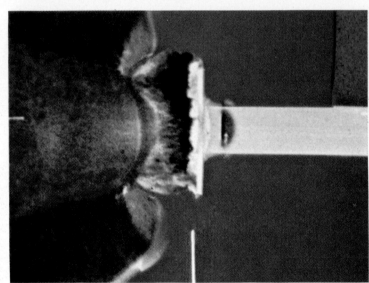

6. After a slight delay, a detonation shows up in the second pentolite stick.

RECORDER OF MICROSECONDS
A camera that can take pictures a microsecond apart was used to photograph the explosion sequence shown above. The lens *(lower left)* points through a bulletproof porthole into the blast chamber and focuses light from the explosion on a revolving mirror, located at the apex *(lower right)* of the pie-shaped camera. As the mirror spins around, it sweeps a reflection of the explosion scene across the film, through the arrangement of slots fanning out about it.

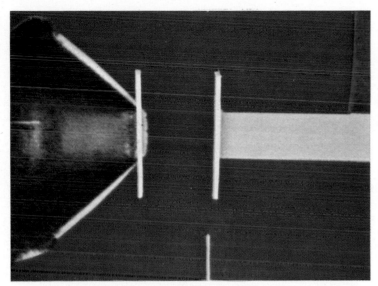

3. The shock reaches a vertical plaster shield at the end of the first stick.

4. The shield holds back smoke as the blast arcs toward another stick.

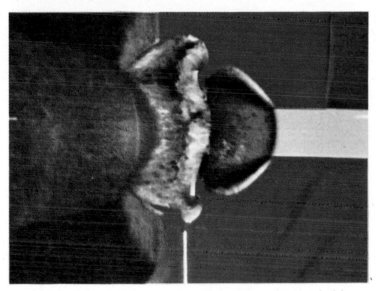

7. New shock waves form as the explosion travels along the second stick.

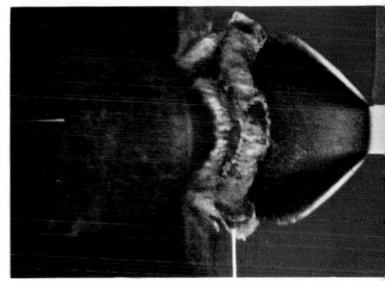

8. The second stick disappears as gases and debris enshroud the explosion.

MILLIONTHS OF A SECOND
The Profile of an Explosion

Explosions take place so rapidly it would seem inconceivable that anyone could time one. A charge of dynamite or TNT blows up in microseconds, or millionths of a second. Yet clocking detonation speeds of such explosives is routine for engineers, who are continually searching for more effective blasting materials.

The very speed of an explosion is the quality that gives it usefulness. When TNT detonates, it undergoes an internal chemical reaction similar to burning—but the reaction is so rapid that the resulting gases expand more quickly than the speed of sound. They form powerful, fast-moving shock waves, similar to the waves that create a sonic boom when a jet plane travels faster than sound. To investigate explosions in detail, scientists rely on an elaborate camera with special lenses and whirling mirrors *(left)*. The resulting photographs, by revealing the timing of a blast, indicate whether it will gener-

ate the shattering effect needed by the military or the slower explosions desired by mining engineers.

The pictures in the sequence shown here depict three-microsecond intervals in the detonation of pentolite, a mixture of TNT and a fast-acting material called PETN used as the explosive charge in torpedoes and some bombs during World War II. They render an instant-by-instant account of a detonation that, from start to finish, took only some 24 microseconds.

BILLIONTHS OF A SECOND
Timing Subatomic Particles

Some of the briefest times ever measured occur in the mysterious realms of nuclear physics. Certain subatomic particles travel at nearly the speed of light, and some are so short-lived that their entire history lasts only billionths of a second.

The protons and neutrons that make up the bulk of the atom's nucleus—and the much lighter mesons that bind the nucleus together—are far too small to be seen by the most powerful microscope. To study their life-span, scientists rely on a complex of pipes, magnets and pressure tanks which makes their movements visible. In the 80-inch (203-cm) bubble chamber at Long Island's Brookhaven National Laboratory

(right), subatomic particles are shot into a tank of liquid hydrogen, held under pressure at the verge of boiling. The particles leave wakes of tiny bubbles in the hydrogen. By studying the curvature and length of these wakes, scientists identify the particles and time the intervals between their birth and disintegration.

Certain mesons are too short-lived even for the bubble chamber to show. Their life spans have been calculated at 10 millionths of a billionth of a billionth of a second—the amount of time it takes light to travel from one side of a proton to the other. Such minute periods, some scientists have theorized, may be the smallest intervals of time man will ever encounter.

SUBATOMIC TRACKS
The lines and spirals in this bubble chamber photograph show the paths taken by speeding particles through liquid hydrogen. The arrow indicates the fleeting trajectory of a K-meson, formed when a pi-meson collided with a proton. The particle broke up after only .4 billionth of a second. A remnant, another pi-meson, then made the large spiral wake in the center.

THE GIANT TRACKER
Pencil poised above the branching tracks left by a proton-meson collision, a technician at Brookhaven indicates the results of a bubble chamber experiment. Behind is the chamber itself, a six-million-dollar instrument weighing 480 tons (432 t). The technician in the center of the middle tier is installing a camera in one of the viewing ports of the bubble chamber.

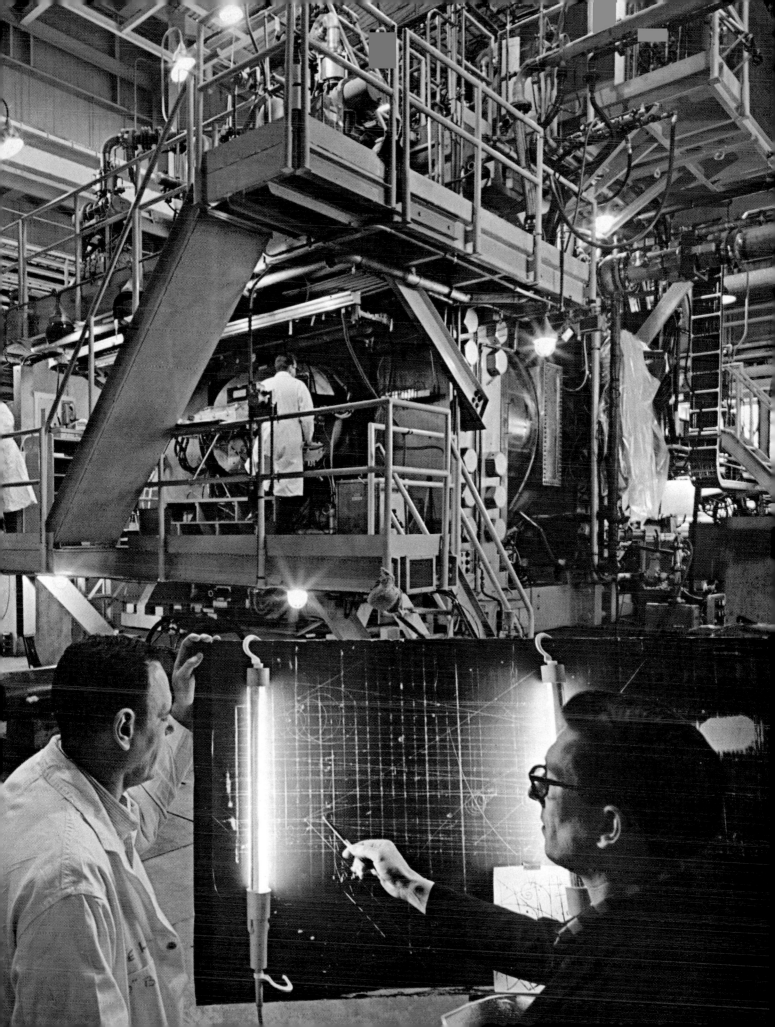

6
Fixing the Start of "Long Ago"

A Neolithic cave painting found in the Sahara documents the immense changes that may occur with the passage of time. The prehistoric artist showed cattle grazing in an area that is now desert but must once have been a verdant pastureland.

FOR SOME 5,000 YEARS, time-reckoners have dealt in ever smaller units. The "natural" unit of the day has been divided into hours, minutes and seconds and the seconds subdivided into milli-, micro-, nano- and pico-seconds. Over the past few centuries, however, this sharpening focus on what one physicist has called "small time" has been paralleled by a growing preoccupation with "big time"—events lying so far in the past that for their timing even a year, let alone a day or a second, is ludicrously inadequate. The archeologist dating a prehistoric campsite measures by thousands or tens of thousands of years; the paleontologist studying the fossil of one of man's fish ancestors, the geologist measuring the age of a mountain range, reckon the years by scores of millions.

Fortunately, nature provides many devices for measuring big time. Layers of rock, fossils, radioactive elements and the dim light of receding stars record the passage of millions and even billions of years. With these "big clocks," men can date the remotest events—calculating the age of the earth and even estimating when the universe itself may have begun.

Even rough reckoning of big time, however, is a relatively recent technique, beginning less than 300 years ago. Practical difficulties helped delay it; events whose time scale transcends the lifetime not merely of a man but of a civilization could not be measured by calendars or ordinary clocks. Even more important was the conceptual lack. Time and change—the daily and seasonal changes in the heavens, the waxing or waning metabolism of a plant or animal, the shifting hands of a clock— are inextricably linked. No man will speculate about measuring time by millions of years unless he believes that changes on that scale exist.

Such an idea was inconceivable in Western civilization as the age of science began. To philosopher and peasant alike, the world had been created at a specific instant, complete and finished, subject to no detectable alteration thereafter. The sun and stars, the earth and all its creatures, retained the same forms they had been given at the Beginning.

Before scientists could seriously set about measuring the extent of the past, they had to reject this traditional view of an unchanging nature. They had to learn to think of past events as a series of continuous changes that, in their own leisurely manner, have created the present. The measurement of big time has therefore formed part of a much broader revolution in scientific thought, which one authority has called "the discovery of time": the recognition that everything in nature changes with time and that the proper concern of science is the study of change itself. The revolution has taken place amidst the dismay, and often against the bitter opposition, of many religiously orthodox people, to whom rejection of the traditional concept of the Beginning has seemed a rejection of morality and of God Himself.

Like any revolution, this one had roots stretching deep into the past. For practical purposes, however, the time revolution in science did not begin until the mid-18th Century. It was then that a brilliant French naturalist, amateur of science and a prolific writer, Georges-Louis Leclerc, Comte de Buffon, set about trying to estimate the age of the

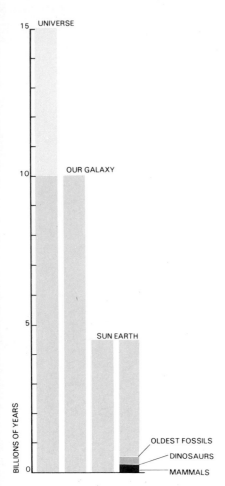

THE IMMENSITY OF TIME PAST is indicated on this graph. The oldest known event, the beginning of the universe *(left bar)*, is thought to have occurred at least 10 and perhaps 15 billion years ago. The Milky Way galaxy, of which the sun and earth are part, was formed about 10 billion years ago, while the earth and sun *(right)* are about four and a half billion years old. By comparison, life on earth is very recent, as shown by the divisions on the earth bar. The oldest abundant fossil remains date back about 600 million years. Dinosaurs disappeared abruptly 70 million years ago; mammals first appeared about 185 million years ago. Man, too recent to indicate on this chart, is a mere 1.7 million years old.

earth. Buffon had already published a book portraying the earth as a molten, luminous ball, torn from the sun by the gravitational attraction of a passing comet. He now attempted to determine how long this white-hot lump would have taken to cool. He heated spheres of different materials and sizes and measured how quickly they cooled. From these experiments he concluded that a sphere the size of the earth would have reached its present temperature in 74,047 years. This figure he adjusted upward, to allow for the heat which the earth had received from the sun, for a grand total of "approximately" 74,832 years.

With today's hindsight, we know that this figure is far too small—about 1/60,000 of the correct value. Among other errors, Buffon grossly underestimated the effects of solar heat on the earth. But erroneous as his figure was, it was still a *scientific* estimate of the earth's age. "His calculations," says a recent commentator, "had proved the essential point: that the time barrier could be breached. By invoking the laws governing familiar physical processes, such as cooling, one might infer the former state of things from the present face of Nature, and determine the dates of physical events far earlier than the first human records."

Reading the earth's archives

Buffon had done much more than this. With a clarity never to be surpassed, he had defined the task of those who were to follow him in seeking to reconstruct the history of the earth and the creatures on it. "In civil history," he wrote, "we consult [documents], study medallions and decipher ancient inscriptions in order to ... fix the dates of moral events." Similarly, in natural history, "one must dig through the archives of the world, extract ancient relics from the bowels of the earth, gather together their fragments, [as] indications of the physical changes which can carry us back to the different Ages of Nature. This is the only way ... of placing a number of milestones on the eternal path of time."

The task of unscrambling the earth's archives occupied many men, but it was the Scottish physician-farmer James Hutton who first explained the geologic processes that had shaped the face of the earth. In doing so, he gave some indication as to how big big time was. Hutton came to geology in a roundabout way. Though drawn to science in his teens, he first "took to the law" as a clerk-apprentice to an Edinburgh attorney. Soon, however, his employer found him "amusing himself and his fellow apprentices with chemical experiments when he should have been copying papers." As a result, Hutton sought a more congenial occupation. His next try, medicine, proved hardly more engrossing, and after a few years of it he became a gentleman farmer.

It is not difficult even now to imagine Hutton striding about his lands, peering into ditches and stream beds to see what lay beneath the soil. Surveying a plowed field in spring, he must have seen raindrops washing earth off the ridges and depositing it in the furrows. Strolling along the nearby seashore, he no doubt observed ripple marks in the bottom of a sandy pool and compared them with marks on a chunk of sandstone cliff torn loose by the waves of a winter gale. As he studied the rock layers, Hutton could see that they had been deposited one on top of the other in an orderly sequence. The bottom layer, he reasoned, must therefore be the oldest layer, and the one on top must be the youngest. Each layer of rock made up a separate chapter in the geologic history of the whole rock formation, and Hutton knew from watching the slow processes that were currently occurring all around him that each layer represented a large chunk of time.

The key to the globe's great mystery

From his observations, Hutton derived what is probably the greatest single generalization in the earth sciences: "The past history of our globe must be explained by what is seen to be happening now." According to this "uniformitarian" doctrine, the hills and mountains, like the plowed ridges, are being continually eroded away by water and wind. The eroded sediments, deposited in layers, are eventually compressed into rock, even as the wave-marked sandstone must have been formed from the rippled sand of some prehistoric beach. Finally, the rocks are slowly pushed aloft, by forces within the earth, to be eroded in their turn. "This earth," said Hutton, "like the body of an animal, is wasted at the same time that it is repaired. . . . Destroyed in one part . . . it is renewed in another."

The cycle of erosion, depositing of sediments, uplift and renewed erosion must, he believed, have occurred repeatedly over a period of time "indefinite in length." With true Scots caution, he refused to estimate its duration; geologic changes, he said, are too slow to be measured by man: "It is in vain to attempt to measure a quantity which escapes our notice and which history cannot ascertain." Instead, he concluded only that "we find no vestige of a beginning, no prospect of an end."

While Hutton grasped the long-term geologic processes that produced a particular rock formation, his understanding of the scope of those processes was limited. He could reckon the time sequence in which the rock layers of a particular formation, say in a cliff on the Suffolk coast, were laid down, but he knew no way of relating those layers to similar ones on a mountain in Wales. It remained for William Smith, a young English contemporary of Hutton, to turn that trick.

The contrast between Hutton, the well-to-do Scottish physician-farmer, and the self-made Smith could scarcely have been sharper. Born to a poor family and left fatherless at the age of seven, Smith laboriously taught himself surveying. For most of his life he made a living as a surveyor and as what would now be called a civil engineer.

Smith's surveying trips around England gave him a perfect opportunity to indulge an interest in rocks and fossils. Recalling in later life one of those journeys, he describes how "my eager eyes were never idle a moment. . . . In the more confined views, where the roads commonly climb to the summits . . . the slow driving up the steep hills afforded me distinct views of the nature of the rocks; rushy pastures on the slopes of the hills, the rivulets and kind of trees, all aided in defining the inter-

THE ORIGIN OF TIME in Chinese legend involves the myth of P'an Ku, who created the universe by chiseling planets and stars out of a cliff representing chaos. The ancestral animals, the Dragon, Tortoise and Phoenix, accompanied him while he worked. When he died, his body gave rise to the physical features of the world—his head the mountains, his flesh soil, his blood rivers, his breath wind. Human beings supposedly descended from the insects which crawled over his body. According to this legend, the exhausting task of creation took 18,000 years, thus establishing a starting point for time.

mediate clays; and while occasionally walking to see bridges, locks and other works . . . more particular observation could be made."

From his particular observations, Smith conceived a solution to correlating a fundamental problem in geology—how to relate the rocks in Suffolk to those in Wales. Pushing his surveyor's traverses across the land and tracing the outcropping of one layer after another, Smith discovered that strata can be identified by the fossils of long-dead animals and plants that they contain. "Each stratum," he wrote, "contained organized fossils peculiar to itself, and might, in cases otherwise doubtful, be recognized and discriminated from others like it . . . by examination of them." By Smith's new technique, fossil-bearing rocks anywhere in England (ultimately, anywhere in the world) could be placed in their proper chronological sequence. Particular rock layers could be related to distant formations, and if nature mixed up the rocks, the relative age of the layers could be reckoned by the fossils they bore.

Beyond this, "Strata" Smith's studies of fossils pointed up a fact of which both geologists and zoologists were already becoming aware: Any rational history of the earth would have to encompass a history of life. If the earth had changed with time, so, it seemed, had living creatures—as witness the strange and sometimes monstrous remains that were, with increasing frequency, turning up in rocks, gravel pits and swamps.

Darwin's monumental contribution

The man who at last mapped out the history of life and explained the natural processes by which it had changed and evolved in time was, of course, Charles Darwin, whose *Origin of Species*, published in 1859, made evolution a household word (in some pious households, almost a dirty word). Darwin did not discover evolution—all the basic ideas had been put forward years before by other men, among them Darwin's own grandfather, Erasmus Darwin. But if the theory that Darwin put forward was less than original, his proof of it was wholly his own—and masterly. In more than 20 years of labor, he put together a solid, reasoned, factual foundation for evolution. He showed that the present forms of living organisms had evolved from other forms in a slow but orderly process—one that must have occupied great periods of time. With Darwin big time became a concept that could no longer be ignored.

His theory withstood the most outraged blasts of the theological traditionalists, but for a while it seemed that scientific objections might succeed in destroying evolutionism where religious objections had failed. Starting from the few known facts about the rate of evolution in living animals, Darwin had guessed that the whole evolutionary process must have taken some hundreds of millions of years. A powerful dissent was filed by William Thomson, Lord Kelvin, one of the greatest scientists of

A RECORD OF TIME PAST, this geological profile of the 50 miles (80 km) between London and Oxford is from one of William Smith's pioneering strata maps. Smith recognized that sediment and fossils are deposited together in layers *(diagonal lines)* characteristic of their time. The differences between fossils in different layers confirmed that these layers were laid down in sequence, with the oldest at the bottom and the youngest on top.

the period. A century after Buffon had sought to estimate the age of this earth from the rate at which it cooled, Thomson posed himself the same problem. With the advantage of a hundred years' advance in the knowledge of the physics of heat, his figure was far greater than Buffon's—but much too small for Darwin. The sun, Thomson calculated, could not have existed for more than 500 million years, while the earth could not have been cool enough to support life for much more than a few million.

At this point there was no reliable way to resolve the conflict between Darwin and Thomson. The "absolute" dating of rocks and fossils was still a matter of educated guesswork. "Relative" dating—putting fossils in their chronological order—was reasonably simple, thanks to the methods applied by "Strata" Smith. From the sequences of layers, it was easy to ascertain that dinosaurs had appeared before mammals, and amphibians before dinosaurs, and even to work out which species of dinosaurs had appeared in what order. But to match layers with years—to determine when dinosaurs appeared, or disappeared—was another story.

Most early methods of absolute dating derived from Hutton's uniformitarian doctrine: The present is the key to the past. By measuring the rate at which a geologic process is taking place today, we can infer how long similar processes required in earlier epochs. For example, if a stream at the bottom of a 200-foot (61-m) gorge is cutting into its bed at the rate of one inch (2.5 cm) in 10 years, we can assume, other things being equal, that it began carving the gorge 24,000 ($200 \times 10 \times 12$) years ago. Or by measuring the average annual rate at which a flooding river piles silt onto its floodplain, we can estimate (other things again being equal) the rate at which similar sediments accumulated long ago.

Though this system provided rough estimates of the ages of geologic strata, it was inadequate to resolve the Darwin-Thomson controversy. Estimates made by this method varied greatly in their reliability. Some eventually proved to be remarkably accurate; others turned out to be far off, and there was then no way of telling which was right.

A big clock discovered by chance

It was not long, however, before an accurate method of absolute dating emerged. About a dozen years after Darwin's death, the French physicist Henri Becquerel left a lump of a salt of uranium on a photographic plate and thereby discovered radioactivity. The marks on the plate revealed that certain elements "decay." As the nuclei of their atoms throw off particles, they transform themselves into other elements. This decay occurs at random among individual atoms, but when great numbers of atoms are averaged, the decay maintains a constant rate. Thus the decay rate becomes much like the ticking of a big clock. It is safe to say that no single discovery has done more to revolutionize

MEMENTOES FROM BYGONE AGES, these fossils *(below)* were among the 13 distinct varieties noted by William Smith in a stratum called London clay *(far right of map, opposite).* All but the shark's tooth are mollusks; the presence of these marine animals in the stratum indicated that this area was underwater when these creatures became embedded in the sediment 40 million or more years ago.

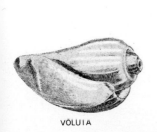

VOLUTA   CERITHIUM   SHARK'S TOOTH   ATHLETA   CHAMA   CALYPTRAEA

man's understanding of time. For geologists and biologists, radioactive atoms in earth, air and water have provided "clocks" by which the ages of rocks and fossils can be accurately measured. Radioactivity provided the first clue to the thermonuclear reactions which have kept the sun burning for at least 10 times as long as Thomson calculated and will stoke its fires for billions of years to come. An understanding of radioactivity has done even more than this. It has revealed that change, by which we measure time, encompasses everything. The stars, whose seeming eternal fixity guided Sumerian calendar makers and the Egyptian hour-watchers, undergo their own birth, evolution and death, even as do the earth and its creatures. The atoms that make up stars, earth and living things are not immutable; formed by processes we can as yet only guess at, they will, it seems, in time be transformed into something else, even if it takes a billion or a billion billion years. The discovery of radioactivity opened the gates to time universal and time illimitable.

How atoms tick off millennia

While Becquerel discovered radioactivity, its potential as a clock was first grasped by the English physicist Lord Rutherford, in 1907. By then, it was becoming clear that the rate of radioactive decay was an almost unbelievably constant process. This fixed average rate is measured in terms of half-life. In one half-life, half of the original atoms decay; in a second half-life, half of what remains, or one quarter of the original; in a third half-life, half of the remaining quarter and so on. Measurements of the decay of ordinary uranium, for example, gave it a half-life of about 4.5 billion years. This decay occurs at a rate of 15 billionths of a per cent per year.

Rutherford knew that every uranium atom, when it decays into a stable atom (of lead), leaves behind eight helium atoms. This should mean that by measuring the amounts of uranium and helium in a rock and calculating the ratio between them, he could tell how much uranium had decayed since it was trapped in the crystallizing rock. Knowing the decay rate of uranium, he could therefore tell how old the rock was.

The theory seemed simple enough, but the practice was more complicated. Helium atoms, being very light, can "leak" away, even through solid rock, and the estimation of how much had leaked was tricky.

The problem was solved in the 1930s with the discovery of uranium and lead isotopes. Isotopes are forms of an element having the same chemical properties but differing in weight. When a radioactive isotope decays, it becomes an isotope of another element; for example, the uranium isotope U-238 ("ordinary" uranium) decays and produces the lead isotope Pb-206. U-235 (fissionable uranium), with a much shorter half-life, yields a second lead isotope, Pb-207. Thorium 232, often found with uranium, decays into a third lead isotope, Pb-208.

With this new knowledge, geochemists could attack the dating problem in a far more sophisticated and precise way. They could compare the ratio of each isotope to its own lead isotope and cross-check by figuring ratios among the different varieties of lead present in the same rock. A fourth lead isotope, Pb-204, which is not the product of any known decay, remains virtually unchanged and provides a further check as to how much lead was originally in the sample.

The experts turned up other radioactive processes to date rocks with little or no uranium or thorium. Potassium 40 decays (in part) to argon 40, with a half-life of 1.3 billion years; rubidium 87 is half transformed into strontium 87 in 47 billion years. By cross-checking among these different clocks, the geochemists were able to refine their theories and techniques — and to come up with increasingly consistent sets of dates.

Reckoning the earth's age

The earliest rocks thus far definitively dated by radioactive clocks were formed about four billion years ago; the earliest remains of living organisms are about 3.2 billion years old. The earth is even older; various indirect dating methods (involving, among other things, comparisons of isotope ratios in terrestrial minerals with those in certain minerals found in meteors that have landed on earth) indicate that it was formed between 4.5 and five billion years ago. Hutton and Darwin have at last been justified in their estimates of the immense reaches of geologic time necessary for shaping the earth and the evolution of life.

Important as radioisotopes are, they have their limitations. They can be used to date igneous rocks, because they are frozen into them when the rock crystalizes. But they cannot be used to date fossils directly, nor can they be used to date the sedimentary rock in which fossils are found. While some fragments of sedimentary rocks may be radioactive, the rock itself is formed by the slow accumulation of particles of different ages, and thus cannot be dated, except indirectly by dating the igneous rock that sometimes invades it. Moreover, the time scale of these long-period isotopes is coarse. Even the most refined techniques have some limitations, and the dates they yield are accurate only to within about 10 million years. This inaccuracy is of little importance in dating say, the epoch during which the Appalachian Mountains were formed, but it is fatal in dating the period in which *Homo sapiens* and all his works have appeared.

A radioisotope clock (or perhaps one should call it a stopwatch) that does for archeology what uranium, potassium and rubidium have done for geology and paleontology was finally contrived in 1946 by the American chemist Willard F. Libby. He used the radioactive isotope of carbon —carbon 14—with a half-life of around 5,710 years, just about right for a useful archeological clock. In nature C-14 is formed by the action of cosmic rays on nitrogen atoms in the upper atmosphere. As it appears, C-14 combines with oxygen to form carbon dioxide, making up about one billionth of the CO_2 that is normally present in the atmosphere.

Libby reasoned that C-14 would constantly be absorbed by living plants, which take in carbon dioxide and chemically build it into their structures. Once a plant died, however, its C-14 content would be "frozen" like the uranium content of a crystallized rock and would begin to decrease by

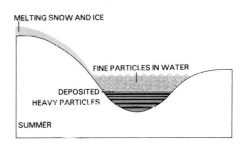

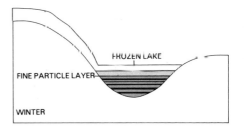

GEOLOGICAL DATING of the past can be accomplished by counting layers of lake-bottom sediment called varves, which are formed as shown here. In the summer *(above)* sediment is carried from melting snow and ice *(blue)* to the lake. The heavier particles settle on the bottom *(gray)*, but wind-caused agitation keeps the finer grains in suspension. In winter *(below)* the lake freezes over, blocking the wind; the fine particles then settle out of the still water and add a thin "icing" *(black)* to the heavy layer. By counting these two-part layers, geologists can determine the age of a varve. Some are as old as 15,000 years.

radioactive decay. By measuring the proportion of the undecayed C-14 to ordinary carbon (C-12), he reasoned, it should be possible to tell how long ago a plant had died. Best of all, since all living organisms contain carbon, the method should be applicable to almost anything that had once formed part of a living creature: wood, charcoal, bone or shell.

For his "trial run," Libby selected materials which could be accurately dated by other methods—tree rings or historical documents. One sample was a piece of wood from the tomb of the Pharaoh Sesostris III, who was known to have been buried about 3,800 years ago. The radiocarbon measurements neatly bracketed this date, giving an age of 3,700 years with a possible error of 400 years either way.

Since Libby's time, C-14 dating has resolved many archeological problems. Similar techniques based upon the decay rates of other atoms have resolved many more. The transformations of potassium into argon, rubidium into strontium, and uranium into lead help date materials which do not contain carbon. As uranium decays, the particles released from the atom leave tiny scars in the material surrounding it. Measuring the profusion of these scars, called "fission tracks," helps scientists pinpoint the material's age even more accurately.

The isotope that will outlive the sun

The study of radioisotopes which has proved so fruitful in measuring changes of big time has led to more fundamental questions about change involving even bigger time. Chief among these is the nature of change within the elements themselves, and whether matter itself may have a beginning and an end. Recent measurements have established that more than a dozen elements have naturally occurring radioisotopes, though they are decaying at rates so leisurely as to make the sluggish uranium atom seem lively. The longevity champion thus far is lead 204. This Methuselah among radioisotopes has a half-life of some 140 million billion years. Long before the earth's supply of this isotope has disappeared, the sun will have burned itself out and the universe as we know it quite possibly will have been transformed into something else.

If the elements (at least some of them) have an end, do they also have a beginning? Apparently they do. Most physicists now believe that the universe originally consisted only of hydrogen. The other elements were synthesized—by nuclear reactions which have been theoretically reconstructed in some detail—out of hydrogen nuclei (protons) and electrons. The synthesis is believed to have taken place in the interiors of stars, which later provided the raw materials for planets. The particular collection of atoms which constitutes the earth and its inhabitants began to be synthesized at a time that can only be conjectured; current estimates range from about seven to 15 billion years ago.

But if the age of the elements can be reckoned, what about the universe itself? Is there a clock to measure its age? Such a measurement— the biggest reckoning of big time—cannot, of course, be made by radioactive clocks; when the universe was formed, radioactive elements did not yet exist. But measurement of the age of the universe can be derived from one of the best-known facts about the universe: its expansion.

The expansion of the universe has been inferred from the so-called red shift. When astronomers examine the light from distant galaxies, they find that its colors have shifted toward the red end of the spectrum. The color of the light depends on the frequency of its waves; thus the frequency of the galaxies' light waves is lower than it should be.

The best explanation of this fact—at least in terms of any physical laws familiar to us—is the so-called Doppler effect. This phenomenon, which affects all kinds of wave motion, was first noticed in sound waves. A person standing next to a railroad track will hear the whistle of a passing train not as a steady note but as a wail. The whistle will drop in pitch (that is, the frequency of its sound waves will diminish) as the train flashes past and begins to move away. In this same manner, the light waves from distant galaxies change, diminishing in frequency. The galaxies, like the train, must be moving away from the observer.

From the red shift we can tell that the galaxies are moving away from us and from each other and how fast they are receding. It turns out the distant galaxies, which appear smallest and dimmest, are receding most rapidly. But where (if anywhere) are they receding from?

The age of the universe itself

Knowing the present position of the galaxies and how fast they are speeding away from us, it is no great trick to calculate how long they have been traveling outward. The results are very curious. It appears that billions of years ago the universe and all the matter in it was crammed together in a volume far smaller than at present. Matter, subjected to almost inconceivable heat and pressure, was not matter as we know it—molecules or even atoms. Even protons and electrons, which we are accustomed to think of as the fundamental stuff of the universe, may not have existed in any form we would recognize.

This intensely compressed universe was unstable and exploded, sending great masses of itself flying off in every direction. The separate masses, expanding and cooling as they flew, became galaxies, part of whose substance eventually condensed into stars and planets. This explosion was the "big bang" which, according to the most widely held theory of cosmology, began the universe. Assuming that the astronomers have estimated the present positions of the galaxies with reasonable accuracy, it must have taken place some 10 to 15 billion years ago.

At this point, the scientific approach to big time reaches its ultimate limit. Science has traced back and dated the origins of men, of life, of the earth, the elements and the universe itself. As to what—if anything—came before that, one can barely guess. In a sense, the problem of big time, which began as a dispute between religious faith and scientific data, has come full circle. As the American physicist William A. Fowler has observed, "Only men of strong convictions, religious or scientific, have the courage to deal with the problem of the creation."

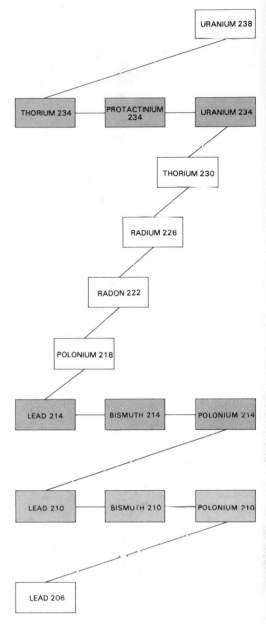

THE SLOW DECAY OF URANIUM reveals the ages of the earth's rocks. A uranium atom *(top box)* of atomic weight 238 first loses a nuclear particle (indicated by vertical drop), and becomes thorium of 234 weight. In the next two steps the atom's nucleus creates and emits electrons *(horizontal lines)* and changes into protactinium 234 and uranium 234 (color indicates near-identical weights). In most cases the process continues as shown until the atom becomes lead weighing 206. Knowing that half of any sample of uranium 238 would change into lead 206 in 4.5 billion years, scientists calculate the age of uranium-containing rocks by determining their proportion of lead 206.

Reaching Backward in Time

The past, beyond a few thousand years of recorded history, once seemed a void, with only an occasional ruin or relic to hint of vanished eras and civilizations. Until fairly recently there was no way of pinning down the authenticity of legendary events said to have occurred "long ago" or "once upon a time." Thus, Homer's epic description of the Trojan War was regarded by many as fiction until a 19th Century archeologist uncovered the long-lost ruins of Troy and identified the gate at Mycenae *(right)* through which King Agamemnon probably led his Greek warriors about 3,000 years ago.

Today, scientists can trace civilizations back through thousands of years, mankind through hundreds of thousands, life through millions, and the universe through billions. No longer content with legends of a vague, dateless past, science has found new clues to ancient events in the growth rings of trees, in the radioactive carbon of organic material and in buried layers of rock. Telescopes have picked up faint glimmers of starlight that originated more than eight billion years ago. And yet, scientists still do not know if the starlight is truly old in terms of the universe—or if its age covers a mere wink of time. Man's probing into past time may be only beginning.

THE GATE AT MYCENAE
"Mycenae, rich in gold," was Homer's description of one of ancient Greece's greatest cities. Abandoned and almost forgotten over the centuries, it was partly excavated in 1876 by the pioneer archeologist Heinrich Schliemann, shown seated *(foreground)* before the Lion Gate to the citadel. Relics found at Mycenae have been dated back as far as 1650 B.C.

Calendars Locked in Wood

When the explorers of the American Southwest first came upon Pueblo Bonito *(right)*, they knew they had found an ancient mystery—but they had no idea just how ancient. Even the Indians who lived in the area could tell nothing about the strange, deserted city, which had once housed 1,200 of their ancestors in a single 800-room building. Finally, in the 1920s, the American scientist Andrew E. Douglass gave Pueblo Bonito a past. He did it with tree rings.

Trees add growth to their trunks by adding rings of cells beneath the bark each year. The age of the tree can be determined by counting the number of concentric rings in a cross-section of the trunk. In many cases these growth rings also indicate the weather of the growing seasons—thick rings for years of heavy rainfall, thin ones for drought. Since generations of trees overlap, patterns can be pieced together *(left)* to form a year-by-year calendar stretching as far back in time as there are trees to fill the gaps.

Near Pueblo Bonito, Douglass took samples of wood from the oldest liv-

READING THE RINGS

In the diagram above, rings from a series of trees are matched to form a year-by-year pattern extending back *(top to bottom)* six generations. The inner rings of each tree match the outer rings of the older tree beneath it. The age of Pueblo Bonito was determined by tracing such a pattern back to its structural timbers *(right)*, cut more than 1,000 years ago.

ing trees he could find. The search for more ancient ring patterns led him to tree stumps and finally to structural beams *(below)* and buried logs in the ruins. By counting backward through the years to the outermost rings of the structural beams, Douglass discovered that Pueblo Bonito was built around 900 A.D.

In his studies of tree rings, Douglass also turned up a possible explanation for the abandonment of Pueblo Bonito in the 11th and 12th Centuries: a series of severe droughts that ruined the agriculture of the area.

Pueblo Bonito, once the center of an agricultural civilization, is now a desert ruin.

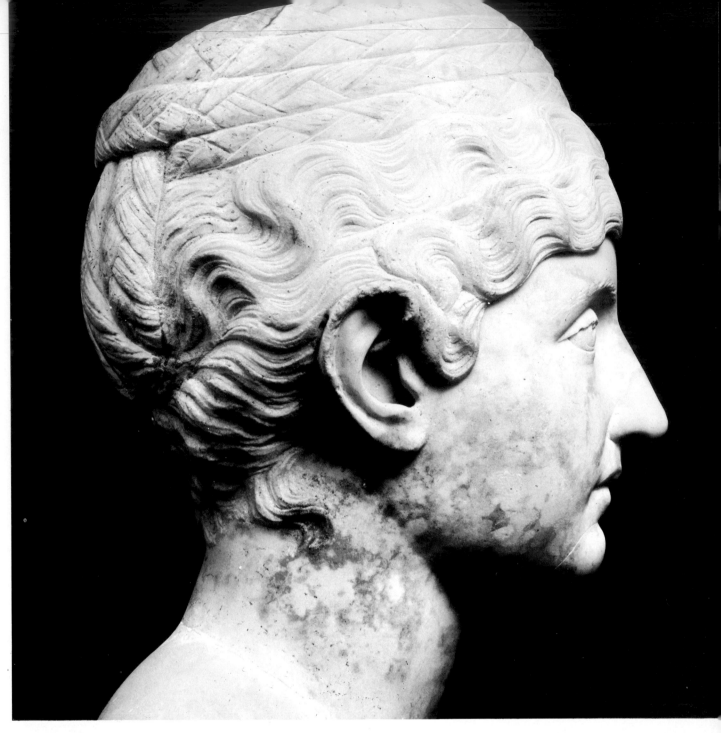

A head of Roman Empress Anna Galeria Faustina *(above)*, who lived in the Second Century A.D., shows a hairstyle that was then in vogue.

Time Clues
in Ancient Trash

The archeologist, digging at the site of an ancient city, prizes the contents of domestic trash dumps as much as he does the rubble of magnificent temples. In these dumps he finds broken tools, weapons, pottery, toys and kitchenware deposited layer upon layer; the deeper he digs, the further back in time he reaches. But the depths of these layers of debris cannot tell him the ages of the civilizations that occupied the city—only the sequence in which they flourished and fell. The clues to the age of each layer are the bits of trivia it contains: a wine bottle of a particular shape, a statuette of a god, a sword, or a wooden doll with a certain type of hairstyle *(right)*. If these objects belong to a specific period in the known history of design, belief or fashion, they can be easily dated.

Shards of pottery, for example, are used to establish the age of Greek

The same style, on the head of a wooden doll *(above and below)* found in a Roman tomb, indicates that it can be dated to Faustina's era.

ruins, because the style of the pottery changed so noticeably from century to century. Vases of the 11th Century B.C. were decorated with fish and plant designs; by the Ninth Century B.C., geometric patterns were preferred; two centuries later, the trend was to Oriental monsters and human figures. Other, more subtle design differences sometimes enable archeologists to date a piece of Greek pottery within a single decade.

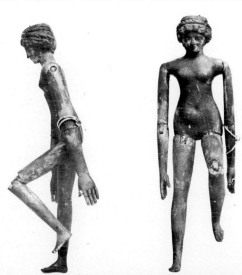

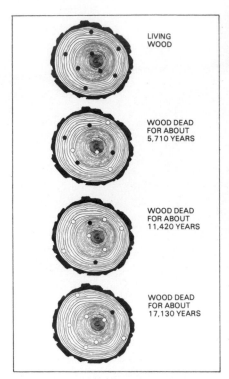

THE SLOW DEATH OF CARBON 14
The radioactive carbon 14 atoms in a piece of wood—shown as black dots in the drawings above—disintegrate slowly at a rate that is known. Each drawing from the top down represents a successive 5,710 year "half-life" —the time it takes for half the atoms to lose an electron and change to nitrogen (white dots).

In the top-to-bottom image labels:
LIVING WOOD
WOOD DEAD FOR ABOUT 5,710 YEARS
WOOD DEAD FOR ABOUT 11,420 YEARS
WOOD DEAD FOR ABOUT 17,130 YEARS

Reading the Carbon Clock

In 1946 the process of telling time backward was made almost as simple as reading a dial. In that year it was discovered that a radioactive clock is ticking away in the bones of mammoths, in the wood of an Egyptian barge (opposite), and in all other material that has lived during the past 50,000 years. The clock is a variety of carbon called carbon 14 (C-14), which tells time by decaying at a slow, fixed rate.

The atmosphere always contains a small percentage of C-14 atoms mixed with ordinary carbon atoms in carbon dioxide gas. Plants absorb the gas and animals eat the plants—each acquiring the same percentage of C-14 and maintaining it in their tissues. After death they no longer replenish this supply, and radioactive decay takes a steady toll of the original supply. Since the wood in a tree trunk (left), is no longer growing, the wood will lose half its C-14 in about 5,710 years, half the remainder in the next 5,710 and so on. In roughly 50,000 years the amount of C-14 left no longer can be detected.

To determine the age of bone, cloth, seeds or seashells—all of which are composed primarily of carbon—scientists first measure the amount of C-14 these objects contain. They must then compare these amounts to the amount contained in some object of known age. Since atmospheric levels of C-14 do change, the standard of reference must be roughly as old as the object to be dated. An ideal standard was found in a venerable tree called the bristlecone pine, which lives for thousands of years. The pine's trunk forms an extremely accurate clock (pages 134–5). After isolating C-14 from wood deep inside the trunks of living pines 9,000 years old, scientists could make a chart correlating the age of each section of trunk with the amount of C-14 the wood—and hence other living things of the same age—contained. Their chart provided an accurate key to the chronology of things that lived at any time during the past 9,000 years.

THE CARBON-DATING TECHNIQUE
In a carbon-dating laboratory, two scientists watch closely as a gas flame incinerates some tiny samples of ancient charcoal inside a sealed tube to convert it into carbon dioxide. After the impurities are removed from the gas, a geiger counter is used to detect pulses of radiation given off by the carbon dioxide's carbon 14. The radiation count, indicating how much carbon 14 remains, is used to calculate the charcoal's age.

THE BARGE OF A PHARAOH
This funeral barge, for Egypt's Pharaoh Sesostris III, provided a decisive test of the new carbon-dating technique in 1949. The carbon clock indicated that the wood was about 3,700 years old—a figure corresponding to the presumed date of Sesostris' death: 1850 B.C.

The Ages of the Earth on File

Spread out before the eyes of every tourist who gapes at the Grand Canyon *(left)* is more than 2 billion years of the past. The record of the earth's history is visible in the layers of rock which were exposed by the Colorado River as it cut a mile (1.6 km) down through a plateau. These layers are composed of different kinds of sedimentary, igneous and metamorphic rock, the successive residues of eons when the Grand Canyon region was covered by inland seas, transformed into a swampy plain, overlaid with desert dunes and finally lifted up by forces within the earth. The youngest layers, near the rim of the canyon, yield fossils that are most like modern animals. Deeper down, the strata hold progressively more primitive forms of life: lizards, shellfish and algae.

Some of the Grand Canyon layers —the igneous and metamorphic rock—can be dated directly by radioactive analysis. However, the sedimentary layers, which contain fossils *(right)*, must be dated indirectly, in terms of their position in relation to dated rock. By matching fossils in one layer thus dated with identical fossils in another undated layer, scientists are able to piece together the puzzles of time.

Although the Grand Canyon provides by far the most spectacular display of strata-history, similar sequential records of the past are available at many points on the planet—beneath the oceans and in the coverings of continents.

A VERTICAL PANORAMA OF TIME
The walls of the Grand Canyon were cut by the Colorado River over a period of at least five million years. The oldest layers at the base of the canyon contain marine animals, such as the crablike trilobites, that dwelled on the bottom of a shallow inland sea 600 million years ago.

EVIDENCES OF ANCIENT LIFE
Digging into strata near Rome, archaeologists unearth the jaw bone of a wild ox from a layer dating back to the Lower Paleolithic Period several hundred thousand years ago. The hole higher up—which held an elephant tusk—was made in a stratum formed thousands of years later.

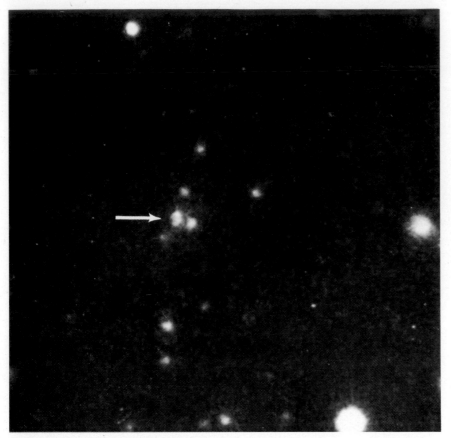

ANCIENT LIGHT FROM FAR AWAY

The speck pointed out in the photograph above is actually a huge group of galaxies, called the Hydra cluster. By spectroscopic analysis *(below, bottom)*, astronomers found that the light of the Hydra cluster of stars took two billion years to reach the earth. Hydra's light is not the oldest known, however. The record-holder so far is a mysterious starlike object called a quasi-stellar radio source, whose light was presumably emitted about 13 billion years ago.

Viewing the Past as It Happens

The astronomer has a unique privilege among scientists: He can actually see the past happening. So vast are the distances of space that starlight—though it travels six trillion miles (9.6 trillion km) a year—often needs billions of years to reach the earth. A galaxy is observed, not as it is now but as it was when its light began the journey to earth.

The antiquity of starlight can be found by determining the distance of the galactic source from the earth. The distance, in turn, is related to the velocity of the galaxies—because all galaxies are moving, and those farthest from earth are apparently moving fastest.

The light itself reveals the speed—hence distance, hence remoteness in time—of the source. The starlight is broken into its spectrum *(left, below)* with a device called a spectrograph, which separates the colors emitted and absorbed by the elements in a star. The colors of many galaxies shift toward the red end of the spectrum, and the amount of shift indicates the velocity of the galaxy. Red shift is gauged by juxtaposing a spectrum of starlight with a spectrum created in a laboratory. The laboratory spectrum shows the colors of the star's elements as they appear when the source is not moving—as they are on earth.

In the spectrum of the huge galaxy shown on the opposite page, the colors have hardly shifted, indicating that the galaxy light is relatively young—about two million years. But the cluster of galaxies shown at left—only a pinpoint in the most powerful telescope—yields a very large shift in the bottom spectrum. Its light as seen today shows how the galaxies looked almost two billion years ago.

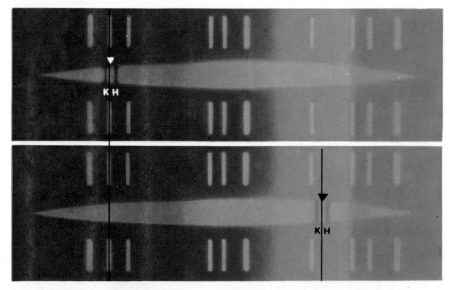

THE EFFECTS OF SPEED ON LIGHT

In the two spectra of the galaxies shown on these pages the starlight is spread against a scale. Both spectra contain dark absorption lines (marked H and K), caused by the element calcium. The H and K lines from galaxy M33 (top spectrum) lie near the blue end, where they would be if caused by stationary calcium on earth. Thus M33 is moving slowly. But in the bottom spectrum, from the Hydra galaxies, the H and K lines have shifted toward the green part of the spectrum, indicating a velocity of about 38,000 miles (61,153 km) per second.

A NEIGHBOR IN SPACE

A galaxy designated M33 by astronomers pinwheels slowly in space, only two million light years away. Seen through a peppering of stars in our own galaxy—the Milky Way—M33 is one of a cluster of about 20 "local" galaxies which display no significant shift toward red.

7
The Einstein Revolution

Albert Einstein, shown here in his Princeton, New Jersey, library, established a new concept of time when he published his theory of relativity. Time, he discovered, is elastic, stretching and shrinking with the motion of the clock that measures it.

TIME IS A PARADOX in that everyone is aware of it yet nobody can define it. This paradox, in turn, stems from an even more basic paradox: the relationship of time to space and motion. Time is measured by motion in space—the motion of the earth about its axis or about the sun, the measured vibrations of a tuning fork or quartz crystal, the rhythmic rotations of a cesium atom. But motion itself is measured by time and space: To describe the motions of a a planet or an atom means to specify the changes in its position with respect to time. Finally, space can be measured by time and motion. The ancient Persians measured distance by the parasang—the amount of level ground a man could cover on foot in an hour. Today we describe stellar distances in light-years—the distance light moves in a year.

This eternal physical triangle, in which time, space and motion are only measurable (and, in a sense, only exist) with respect to one another, has troubled philosophers since St. Augustine first puzzled over the indefinability of time. Physicists, however, have learned to live with the paradox. Indeed, the most fundamental advances in physics have come about precisely through attempts to define these curious relationships more rigorously. Newton's greatest contribution to science was his mathematical definition of how, and under what circumstances, motion changes with time. Later, Einstein revolutionized physics by showing, among other things, how time changes with motion.

Today's physicist, basing himself on both Newton and Einstein, does not see either time or motion as an "absolute," unvarying concept. Whether something is moving, and in what direction, and how long it will take to get wherever it is going, are questions without single "correct" answers. Each answer depends on the individual who gives it and on his particular observation point. A spacecraft, tracked by radar from Cape Kennedy, is seen to be moving away from the earth at seven miles (11.2 km) a second—yet an astronaut in that spaceship sees the earth as moving away from *him* at the same velocity. To the physicist, both views are equally "true." It can be shown, moreover, that to an earthbound observer, the astronaut's clock, no matter how carefully adjusted before launching, will seem to be running slow, and that to the astronaut, clocks on *earth* will seem to be running slow. Again, both observer and astronaut are "right." The simple question, "What time is it?" may be answered differently, depending on who is asking and who is answering, and how they are moving with respect to each other. Even the still more basic question of whether a given event precedes or follows another may have different answers, shaped by the relative motions of the events and of the observer reporting them.

Thus in explaining time, space and motion, scientists have been compelled to move further and further from the "obvious," commonsense view of reality. Yet the uncommon sense they have arrived at, difficult as it often is to comprehend in detail, describes the phenomena of the physical world with an elegant simplicity. The world's first attempts to define time and motion came from the ancient Greeks, who disputed the subject among themselves no less contentiously than they did most other things. Among the most vociferous debaters were members of the so-called Eleatic school of philosophy, who sought to dispose of the time and motion problem by declaring—and logically "proving"—that motion was only an illusion. Their approach, however, was sterile; one cannot dispose of a real problem by providing that it does not exist, and in life, if not in logic, things *do* move and must be coped with on that basis.

Aristotle's time paradox

A more down-to-earth approach, and one that proved far more influential, was that of Aristotle. He was, among other things, probably the first man to state the time-motion paradox. "We apprehend time," he declared, "only when we have marked motion." Yet, he added, "not only do we measure the movement by the time, but also the time by the movement, because they define each other."

When Aristotle tried to explain the nature and cause of motion, however, his commonsense approach proved treacherous. He accepted the apparent "fact" that any moving body has a natural tendency to come to rest. A tossed stone will soon roll to a stop; a cart will cease moving when the horse stops pulling. Evidently, then, nothing moves of itself; something must be moving it.

From this it was but a step to the theory that the velocity of movement was directly proportional to the force, or push, that caused the movement. A two-horse cart will "naturally" move twice as fast as a one-horse cart. A two-pound (0.9-kg) rock, having twice as much "internal force" (that is, weight) as a one-pound (0.45-kg) rock, will naturally fall twice as fast. (Common sense again: Everyone knows that a stone falls faster than a feather.)

Aristotle's mistaken ideas on motion, as on many other subjects, dominated science for centuries after the classical world had declined and fallen. Bit by bit, however, philosophers chipped away at his naive and incomplete view of the world. As early as the 13th Century, some scholars had taken the crucial scientific step—which Aristotle never took—of defining precisely what velocity is. Nobody, after all, can talk about a thing intelligently, let alone measure it, unless he knows what it is he is talking about. The medieval scholars brought time into the definition. A moving object, they saw, was simply changing its position in space. Velocity, then, must be defined as *how much* an object's position changes in a given amount of time. We still express velocity in this way—so many feet per second or so many miles per hour.

The next step came from Galileo, who, among his many other achievements, exploded Aristotle's erroneous views on falling bodies by actually dropping objects of different weights from a height and noting that they reached the ground at the same time. (The story that he performed his experiment from the top of the Leaning Tower of Pisa is probably a legend, however.) Even more important, the great Italian actually *timed* moving bodies. He rolled metal balls down inclined planes of different lengths, and as they rolled he caught water, dripping from a clepsydra, in a cup. By weighing the water very accurately, he was able to obtain precise measures of how much time was required for a ball to roll down the various planes.

From these experiments Galileo realized that velocity alone—change in position with time—was not enough to define motion. It was also necessary to consider acceleration—the change in *velocity* with time. Today, the notion of acceleration is commonplace. A sports car enthusiast—who might be hard put to define acceleration—will nonetheless boast that his machine can go from 0 to 60 miles (0 to 96 km) per hour in 10 seconds. That is, its velocity will increase by six mph (9.6 km/h) every second. Aviators and astronauts measure acceleration in g's—an increase in velocity of 32 feet (9.7 m) per second *per second.*

In the 17th Century, the notion of acceleration was revolutionary. Just how revolutionary became clear when Isaac Newton generalized the observations of Galileo and other scientists into his famous laws of motion.

Newton followed Galileo in rejecting Aristotle's "commonsense" notion that moving bodies "naturally" come to rest. On the contrary, he said, a moving body will never come to rest unless something stops it. A falling rock stops because it hits the earth. A moving cart will roll to a halt because of the friction of its wheels with the road and with the axle they turn on. On a perfectly smooth and level road, with frictionless axle bearings, the cart would roll on forever.

Thus, in the words of the modern British mathematician Hermann Bondi, Newton found that velocity is irrelevant, in the sense that it requires no explanation. What is relevant is acceleration: changes in velocity. Acceleration, said Newton, occurs only when force is exerted on an object—and the amount of acceleration is directly proportional to the force. This simple statement, which for the first time correctly defined the relationship of motion to time and of both to force, was without question one of the greatest scientific discoveries ever made. It led directly to the formulation of Newton's law of universal gravitation in 1684 and indirectly to nearly all the important advances in physics during the next two centuries.

Newton's revolutionary definition

Newton carried his analysis of motion even further. Uniform motion—motion in a straight line, at constant velocity—was, he declared, not absolute but relative. That is, if two bodies are in uniform motion relative to each other, there is no possible way of determining which of them is "in motion" and which "at rest"—except with reference to a third object.

To take a familiar example, most of us have had the experience of sitting in a train in a station and seeing another train, alongside us, begin to move. For a moment, we are uncertain whether our train or the other one is moving—until we look out the opposite window at the motionless platform. We then realize, of course, that our train, which we thought was moving, is motionless relative to the station while the other train is not.

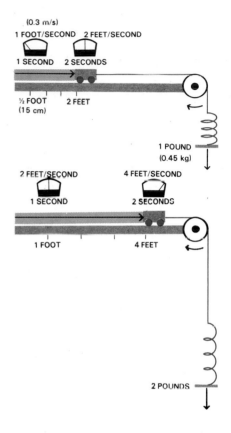

RELATING MOTION TO TIME, Newton formulated laws which became fundamentals of physics. In his famous second law, illustrated here, he demonstrated that force (the pull of a spring) is equal to mass (represented by the cart) multiplied by acceleration (the rate at which speed increases with time). In the first example *(top)*, a force of one pound (0.4 kg) increases the speed of the cart by one foot (0.3 m) per second every second. If the pull is increased to two pounds (0.9 kg) *(bottom)*, the cart accelerates two feet (0.6 m) per second every second. Newton's laws were based on the idea that time, by which we measure motion, is an absolute quantity. This idea holds up well except when motion approaches the speed of light. Then, as Albert Einstein later showed, time slows and appears to be almost standing still.

Suppose that our train is traveling at constant speed, on a straight, smooth track. We can tell the train is moving by the noise or by looking out the window. But so far as our own motions are concerned, it might as well be standing still. As we walk along the aisle we must watch our footing only when the train speeds up, slows down or changes direction. Velocity is purely relative; only acceleration is absolute.

For motion, then, Newton had formulated a "theory of relativity"—though of course he did not call it that. But if uniform motion is relative, not absolute, is not time, which is measured by motion, relative as well? Newton refused to accept this conclusion. "All motions may be accelerated or retarded," he wrote, "but the flowing of absolute time is not liable to any change." His view wrenched time away from its association with motion, for if motion is not absolute, how is absolute time to be measured or perceived?

Newton thought he could prove the existence of absolute time by means of another form of motion which, he believed, *was* absolute—rotational motion. If a pail of water, he said, is hung on a twisted rope and allowed to rotate as the rope untwists, the water will gradually acquire the motion of the pail. In so doing, the water will "ascend to the sides of the vessel, forming itself into a concave figure . . . and the swifter the motion becomes, the higher will the water rise," Newton said. The height to which the water rose, then would provide an absolute measure of its—and the pail's—motion, and, in principle, an absolute measure of time as well.

To clarify Newton's point, let us suppose that a tiny man is standing on the rim of the moving pail. From his vantage point he can see the pail, the water and the ground. But if he looks first at the pail and then at the ground, he cannot tell which is moving. Only when he looks at the water and sees it rising against the sides of the pail—an action which could only take place if the water were moving—can he tell that the water and the pail are moving and that the ground is still.

The fallacy of "absolute" motion

After Newton's time, later thinkers revealed the flaw in this reasoning. The pail, they pointed out, was rotating in the material universe and rotating relative to the objects in the universe. Newton had assumed that the experiment would give the same result *if the pail were rotating in empty space*. But there was—and is—no way of *proving* this assumption experimentally, and therefore no proof that the pail's motion relative to the universe is in fact absolute.

This rather involved philosophic argument, while it more or less disposed of absolute time as "not proven," did not establish the relativity of time in any but the most abstract sense. The discovery of the true relativity of time, and its relationship to motion, did not come until the dawn of the 20th Century. Like many important scientific discoveries, the modern theory of relativity got its first impetus from experiments which seemingly were concerned with an unrelated matter. These experiments dealt with the nature of the "ether" and involved very precise measurements of the velocity of light.

Physicists of the 19th Century viewed light as a series of waves, similar to sound waves or to a train of ripples on the surface of a pond. If, as they thought, light was a wave, common sense said that it must be traveling *through* something, just as ripples travel through water or sound waves through air or other substances. Yet light was clearly different from sound. If a physicist sealed a ringing alarm clock in a glass jar, from which he then evacuated the air, he would hear the sound die out as the air thinned out. But he would still see the clock—meaning that light reflected from it was passing through the vacuum to his eyes. Light evidently did not need air to be transmitted.

But if not air, then what? The physicists's answer to this question was the ether. This singular substance, which was held to permeate everything in the universe, was, it seemed, completely weightless. In fact, it had no properties of any kind which physicists could observe. The only positive thing one could say about the ether was that it was the stuff through which light traveled.

Physicists were not happy with the ether theory. It is uncomfortable to have to make statements about a substance whose properties cannot be detected or measured. Nonetheless, no other explanation of light transmission seemed possible.

A universe full of "ether"

In 1887, two American physicists, Albert A. Michelson and Edward W. Morley, thought they had discovered a way to observe the ether. They reasoned that if it pervades the universe, the earth must be moving through the ether on its trips around the sun. And this movement through the "ether stream" would necessarily affect the velocity of light. A light ray shining in the direction of the earth's orbital motion would be traveling "against the stream." It would therefore move a little slower than a ray shining in the opposite direction as it takes a little longer to row a boat upstream and back than to row it across the river and back. The difference in velocity between the two light rays would be minute—about one part in 5,000. But the two physicists had devised an apparatus that could measure far smaller differences.

The experiment, however, showed no difference at all. The velocity of light did not change, whether it was traveling "with" or "against" the earth's motion or, for that matter, in any other direction.

Within 20 years after the Michelson-Morley experiment, a young clerk in the Swiss patent office began some hard thinking about its results, as well as a number of other puzzling facts and theories that physicists had turned up. Albert Einstein realized that one possible conclusion from the experiment was that the ether simply did not exist. An additional conclusion, which he made into a basic assumption of his theorizing, was that the speed of light is always the same, regardless of whether the source of light or the observer are moving relative to each other. This was a triumph of

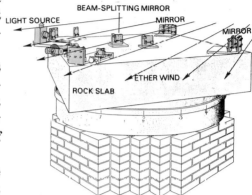

SUPPORT FOR EINSTEIN'S THEORY came from an 1887 experiment with a series of mirrors *(above)*. Albert Michelson and Edward Morley used the apparatus to investigate a theory that light traveled through a substance called ether, which blew around the earth like a wind. To check the effect of ether on the speed of light, they placed the mirrors as shown in the simplified diagram below. A beam of light *(blue-gray)* was split by a semitransparent mirror so that half *(gray)* traveled parallel to the wind *(arrows)* and half *(blue)* an equal distance across it. The physicists found that both halves went at the same speed, thus refuting the ether wind theory. Einstein carried the findings further and came up with a basic postulate of relativity theory—that the speed of light is constant under all conditions.

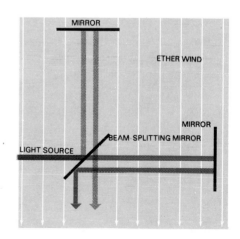

uncommon sense. Ordinarily speeds add or subtract fairly simply. Suppose, for example, two planes are flying toward each other, shooting bullets all the while. It is obvious—and also true—that the bullets will arrive at their targets much more quickly than if the planes were flying away, with their guns firing backward.

In fact, at supersonic speeds the bullets will *never* hit the target in the latter case. Even if a bullet leaves one plane with a relative velocity of several thousand feet per second, the other plane has a still greater relative velocity, and the bullet will never catch up. Einstein saw that light must behave quite differently. Its speed, the Michelson-Morley experiment had shown, is absolute, unaffected by any relative motion. The light from a star 20 light-years from earth will take exactly 20 years to reach us, whether the star is speeding toward the solar system or away from it. If light behaves in this way, Einstein realized, this must mean that time is as relative as motion is. Just as there is no absolute frame of reference by which motion can be measured, so there is no absolute time. If two clocks are moving relative to each other, they will keep different times— and there is no way of saying which clock is "right." If observers are moving with the clocks, each observer will be convinced that the other man's clock is running slow.

An imaginary experiment in relativity

To understand this paradox, let us employ a favorite device of the physicist, a "thought experiment." Such experiments are performed not in the laboratory but in our imaginations. The only requirement is that they must not contradict any of the known laws of physics.

Suppose, then, that we have two enormous spaceships, each about 500 feet (152 m) in diameter and made of tough, transparent plastic, lying side by side somewhere in space. As shown in the accompanying drawing, each has a sort of "master clock" aboard. This clock consists of three parts: on one wall inside of the ship, a flashing light and a screen, set very close together; on the opposite wall, a mirror. The mirror is so arranged that it will pick up a light flash from the other wall of the ship and reflect it back across the ship to the screen. The spacing is such that the light will take exactly one microsecond to make its journey across the ship and back from light source to screen.

Because of the transparent construction, an astronaut aboard either spaceship has a clear view of what is happening inside the other ship. He can see its light go on and then, a microsecond later, see a blotch of light appear on the screen beside it. By an appropriate system of time signals, he can adjust his own clock to the other ship's so that they are precisely in step. But let us now suppose that the two ships—call them A and B—are moving past each other at about one third the speed of light.

Astronaut A watches his light beam making its 1,000-foot (305-m) journey across his ship and back and observes that it still takes one microsecond. But when he looks at ship B, things have changed.

The light beam aboard B will still be traveling from source to mirror to screen. But its path—as seen from A—will no longer be 1,000 feet. In the time that the beam takes to travel back and forth, ship B, together with its light source, mirror and screen, will have moved 333 feet (101 m) relative to ship A. Thus B's light beam, seen from A, will have traveled more than 1,000 feet. Because ship B is moving relative to ship A, its beam of light—as seen from ship A—will also move forward as it crosses and recrosses the ship. Instead of traveling from its source to the mirror and back to the screen at right angles, it will follow an oblique path each way. This means that the light—as seen from ship A—will travel slightly more than 500 feet (152 m) in each direction. By using geometry it is possible for us to calculate that the total distance of this greater path is 1,054 feet (321 m).

Since the speed of light is not affected by the motion of its source, it follows that B's light beam, as timed by A's clock, must take more than one microsecond (actually 1.054 microseconds) to travel the "longer" path from source to screen. From A's standpoint, therefore, B's clock will be running slow. And of course if B looks at A's light beam—which will be moving relative to B's ship—he will see exactly the same thing. To him, A's clock will also be running slow. And the truth is, of course, that both astronauts are right.

The proof of Einstein's theory

At this point, a certain amount of skepticism may be in order, if we recall, for example, Lord Kelvin's "proof" that the sun and earth could be no more than half a billion years old. Thought experiments, no matter how ingenious, do not really prove anything. If Einstein's theories had been unprovable by observation, they would have remained fascinating but useless constructions, as was Newton's theory of absolute rotational motion. But the crucial fact is that observation in and out of the laboratory has proved Einstein right.

Though we have no spaceships that can travel at anything like one third the speed of light, we do have great machines—atom smashers—which can accelerate atomic particles to velocities approaching the speed of light. Indeed, some "natural" particles, formed in the upper atmosphere by cosmic rays, move at speeds comparable to that of light. All these particles, like the radioisotopes used for dating, are unstable and can therefore, like the isotopes, be used as clocks. Their half-lives, by which scientists reckon their life expectancy, are measured not in years but in microseconds. The "time expansion" which we observed in our

THE SLOWING OF TIME at high speeds, one tenet of Einstein's relativity theory, is illustrated by the two imaginary spaceships in this drawing, moving past each other at one third the speed of light. Each ship contains an identical clock like that diagramed on ship A, consisting of a light mounted on one wall emitting flashes which are reflected by a mirror on the opposite wall and returned to a screen next to the flasher in one microsecond *(dotted lines below)*. Both clocks were synchronized before the ships were launched, but as the observers in ship A watch B moving past them, they notice that the light in B's clock takes longer to reach the screen than the light in their own clock. The slowing is a result of B's relative motion; from A's point of view, B's light no longer travels straight back and forth but takes a longer oblique path *(dotted lines above)* as B moves forward. Since the speed of light is constant regardless of the motion of its source, it takes B's light longer than a microsecond to travel the increased distance, and thus B's clock seems to be running slow.

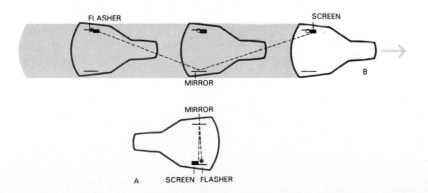

FLASHER SCREEN

MIRROR

B

MIRROR

A SCREEN FLASHER

imaginary spaceship experiment was first detected in real life during studies of the cosmic-ray particles called muons. Produced high in the atmosphere, they hurtle toward the earth at a speed close to the velocity of light. But since they have a half-life of only about two microseconds, they thin out rapidly as they travel downward.

It turned out in actual experiments, however, that the muons were not thinning out as rapidly as their normal half-lives would indicate. When the particles were "counted" on the top of a mountain and then in the valley thousands of feet below, there were something like six times "too many" muons that survived long enough to reach the lower altitude. The only explanation was that their half-lives were longer at high speeds—that is, their clocks were running slow. This interpretation has been confirmed in the laboratory. The lifetimes of fast-moving unstable particles have been precisely measured. In every case, the lifetimes of the particles increase as the speed goes up—and the amount of increase is just what Einstein's equations predict.

Making men age more slowly

Space travel at extremely high speeds could produce precisely this sort of "time-slowing" in human beings. An astronaut might kiss his newborn son goodbye, travel through space at a very high speed for five years with his clock "running slower" than clocks on earth and return to find the boy in high school. At first glance, this may look like a recipe for near-immortality, since the astronaut would have aged only five years during 15 "earth" years. But though he would live longer relative to his family on earth, he would not do so relative to himself. So far as his own biology and consciousness were concerned, he would age at a normal rate and live no longer than his appointed span.

As a practical matter, paradoxes of this sort are not likely to occur soon, if ever. In our spaceship thought experiment, a relative velocity of one third the velocity of light "stretched" time by only about 6 per cent. And to make a real spaceship travel at such a speed would require billions of times the energy now used to put a satellite into orbit.

But while relativistic effects become significant only at very high speeds, they can be detected in the laboratory at comparatively low speeds, by using the frequency of atomic radiations as clocks. Until fairly recently, these frequencies could not be controlled with sufficient accuracy to permit such experiments. But in 1958, the German physicist Rudolf L. Mössbauer showed that the frequency emitted by certain atoms that are rigidly bound in a crystal stays constant within extremely narrow limits. Mössbauer's discovery, which won him a Nobel Prize, suggested to a group of British physicists another experimental test of relativity. The apparatus they used was something like an atomic clock crossed with a phonograph. At the center of the turntable, they mounted a crystal containing "excited" atoms, which emitted gamma rays (ultrashort X-rays) of a very stable frequency. On the edge of the turntable they placed a crystal containing similar but "unexcited" atoms that would greedily absorb the gamma rays. The degree of absorption of the gamma rays could be measured by a detector.

When the turntable revolved slowly, the rays were mostly absorbed. When it was speeded up to about 500 revolutions per second, however, the detector showed decreased absorption. Evidently, the atoms of the absorber had begun to react to a different frequency and could not absorb the radiation. That is, they were now operating on a different time scale from the atoms at the center of the turntable because they were revolving around it.

One by one, most of the seemingly illogical assertions of the theory of relativity have been proven in real laboratory experiments. One of the theory's central principles, that time is deformed not only by very high speeds but also by strong gravitational fields, was confirmed in 1976. Einstein believed that in the vicinity of a very massive body, such as the earth, clocks run slower than they would in empty space. Large bodies in effect "warp" the continuum of time, he said, just as they warp the trajectories of bodies moving through space by exerting a gravitational attraction on them. To test this assertion, scientists hoisted a hydrogen maser clock—which was accurate to one part in a hundred trillion—aboard a rocket and sent it 7,000 miles (11,265 km) into space. Sure enough, as the rocket bore the clock away from the earth's gravitation it began to run fast: In the course of the trip it gained about 47 nanoseconds relative to an equally accurate clock on earth. Time has proven to be as elastic as a rubber band, just as Einstein foresaw.

When "before" means "after"

The relativity of time encompasses not merely the expansion or contraction of time but even its most basic aspect—its "before" and "after" properties. By extremely complex reasoning it can be shown that for observers moving in different directions or at different speeds, there are certain sequences of events in which there is no absolute past or absolute future. One observer, that is, may find that event A occurred before event B; another, that they occurred in the reverse order; a third, that they were simultaneous. All three are "right"—within their own frames of reference.

The traditional view of time is often symbolized by the two-faced Roman god Janus, who, standing in the present, looks backward to the past and forward to the future. Relativistic time has, in effect, given Janus three faces. One looks toward the "absolute past"—events which any observer, anywhere, would agree took place before his own present. Another looks toward a corresponding "absolute future." The third face looks—in a somewhat cross-eyed manner—at the region of "time indefinite," in which the relationships of past, present and future depend on who is defining them and his own motion. Relativity, indeed, is an excellent example of what the British scientist J.B.S. Haldane meant when he said that the universe "may not only be queerer than we imagine, but queerer than we *can* imagine."

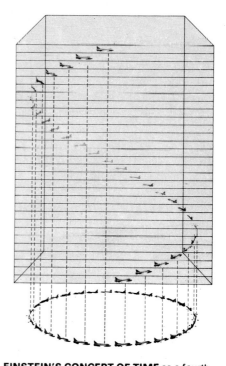

EINSTEIN'S CONCEPT OF TIME as a fourth dimension—essentially equivalent to length, width and depth—is illustrated in simplified form by the diagram above. The pattern of a circling airplane, photographed at one-minute intervals *(bottom)*, has dimensions of length and width. But if the individual photographs are arranged in ascending order *(top)*, the time dimension can be perceived. A relationship has been established between the pictures and the times they were taken. The resulting corkscrew pattern shows what physicists call a three-dimensional continuum and represents the airplane's motion through both space and time. The dimension of depth is omitted here.

The Great Relativity
Bomb Plot

When Albert Einstein advanced his special theory of relativity in 1905, he turned upside down everything that common sense and science had established about time. He said that time is not absolute but is a relative quantity that could show one value to one observer while seeming different to a second viewer. The whole thing seemed preposterous.

Though most scientists came to accept the Einstein theory, its principles have continued to elude many of the less informed—including, to his lasting regret, the villainous fellow at right. He is the sly and ruthless Agent X, an international comic-strip spy who is plotting a daring act of sabotage aboard the most fantastic train in the universe—the Relativity Express. The Express travels at a speed approaching that of light. At this velocity the extraordinary effects of relativity are apparent. Objects shrink in length. Past, present and future become wildly mixed. Moving clocks do not remain synchronized with those standing still; even man himself ages less rapidly. Such astonishing events will cause the downfall of Agent X. Shrewd and cunning he may be, but like all criminals he has committed a fatal error: He has failed to keep himself informed about the odd effects of relativity.

The notorious Agent X, beady eyes hidden behind dark glasses, gestures toward the sleek Relativity Express, the train that is to carry forward his fiendish plot to blow up the world's greatest nuclear power plant. Hidden aboard the express is a bomb, carefully timed to go off as the train passes the plant. At that moment the express will be traveling at 140,000 miles (225,302 km) per second, three quarters the speed of light. X is confident. He has calculated well and is certain his plot will succeed. But he is in for surprises!

 "Everything depends on time," muses X as he sneakily watches the loading of an extremely accurate clock on the Relativity Express. The clock is the key to his dastardly scheme. Inside the clock Agent X has planted his bomb, set to explode when the hands point to 1:30, the very moment the train is scheduled to pass the nuclear power plant. Only this precise clock and this dependable train can achieve X's evil purpose—if either the clock or the train runs just one second slow, the bomb will go off 140,000 miles past target!

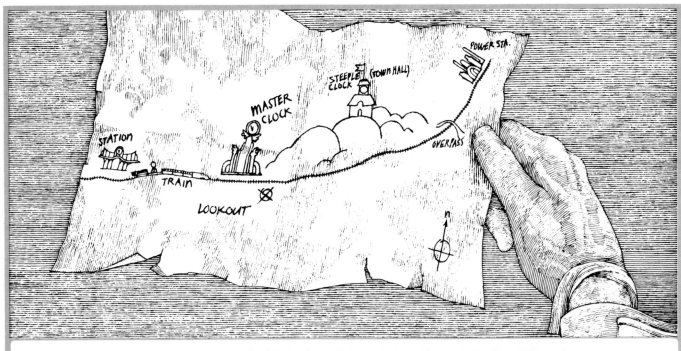

 Agent X has prepared a map of the bomb plot. He knows that the train clock is due to be set in accordance with the giant master clock near the track; this, he assumes, will insure his bomb of split-second accuracy. Just to be sure, however, he plans to check the train clock once again as it passes the steeple clock near the power plant. Meanwhile, his lookout will be watching for the train; once it passes he will proceed in his jet-powered getaway car to a meeting with X at the overpass—as the Express roars toward its rendezvous with doom.

4 "On time!" gloats the cunning agent as the Relativity Express passes by the master clock, at the very moment it sends its noontime signal to synchronize the train clock. Now the bomb must be correctly timed, thinks X smugly, proud of his wicked plot. But little does he reckon with the effect of relativity on time at the train's great speed. The strange shape of the clock tower offers a clue, but in his exultation Agent X fails to notice. The wily spy will suffer severely for his ignorance; he is about to learn about relativity the hard way.

5 But suddenly X sees that something strange is happening! The lookout is a broad-shouldered, heavyset man —but this fellow standing at trackside is incredibly skinny. What's more, the bench and even the trees are peculiar, pressed into narrow shapes. Only the recognition of a hand radio like the one that controls the getaway car indicates to X that this might be his accomplice. He waves frantically at the man as prearranged but he is not sure this is the right chap. His first encounter with relativity has shaken him badly.

6 The lookout, who is really as squat and sinister as Agent X remembers him, is equally stunned by what he sees. He was told the train would be streamlined—but instead it is like a string of squashed-up streetcars. And look —oval wheels! He, too, does not know that relativity makes an object shorter when observed in motion. At last the lookout spots a distorted man waving madly from the rear car. Shaking his head, he prepares to follow in the getaway car. He is not sure who the fellow is, but he proceeds on the hunch that it is his leader, Agent X.

 Beset by doubts, Agent X turns again to his map for reassurance. The next landmark is the town with the steeple clock. "I'll check the clock on the train one last time when we pass that accurate steeple clock," he mutters. "If the two clocks agree—and why shouldn't they?—the bomb will be perfectly timed and the plan will succeed. Then I'll drop off the train." But now the once-cocky undercover man is gripped by fear. What could have made the lookout seem so unfamiliar? Something outlandish is interfering with the plot!

8 The first real evidence that something has indeed gone wrong is spotted by an innocent bystander, a little old lady passing near the steeple clock as the Relativity Express speeds by. The train is actually moving so fast that only in a comic book would it be visible, but the sharp-eyed little old lady can read the clock; it says 12:40. "How odd!" she remarks. "The steeple clock says 1:00 and it's been right ever since I was a little girl. The funny clock on the train must be slow!"

9 At the same moment Agent X spots the steeple clock, leaps from his seat in disbelief and rushes to the car door to check the clock on the flatcar. Sure enough, it says 12:40, although the steeple clock reads 1:00. "Ye gods!" he shouts. "That steeple clock must be wrong! It must be fast—or else it was set wrong! I know the train clock was correct back at the master time station." The whole plan is in danger! In near-panic, X looks for the conductor to tell him that the world has gone mad; the clocks are all wrong.

10 "Both clocks are right," smiles the knowledgeable conductor. "But each clock is running slow relative to the other. As you can see, the steeple clock, which is moving relative to us, has a second hand that ticks more slowly than the second hand of our train clock. But it depends on where you stand: The people on the ground see our second hand as ticking slowly. Furthermore, the steeple clock and the master clock were never really synchronized from our point of view—although they were from the viewpoint of someone on the ground."

11 There is only one hope for the desperate Agent X now: He must reset the terrible bomb to take into account the time change caused by relativity. He has tricked the conductor into telling him that they will pass the power plant when the train clock says 1:00. Now Agent X must change the bomb mechanism to explode then. He slips from the coach onto the flatcar bearing the clock. If only he can reach the clock without rousing the drowsy guard, he may win his battle with relativity after all.

12 Agent X creeps up to the clock stand, against which the guard is resting, slowly opens the door and reaches inside for the bomb. The seconds tick by as Agent X tries to adjust the bomb mechanism. But the noise of Agent X at work nearby awakens the guard, who leaps to his feet. Agent X hastily abandons his work and jumps to the top of the coach to try to make his escape. "Rats!" spits out the villain. "The plot is a failure; I must flee before all is lost." But the guard hangs on to his coattails.

13 As Agent X doffs his coat he shows he is well prepared: Strapped to his back are rockets capable of propelling him at half the speed of light. Away he zooms! "Success!" he exults. "Since the train is traveling at three quarters the velocity of light, I must be going at one and one quarter times the speed of light—enough to escape any gunfire!" But if X really understood relativity he would know that no object can attain the speed of light, much less surpass it. X is going at ten elevenths the speed of light—and that will make a difference!

14 A railway guard with the eyes of an eagle notices the commotion on top of the crack Relativity Express. Since anything can happen in a comic strip, he is able to race up the overpass before the train, which is zipping along at 140,000 miles per second, passes by. His laser pistol is in his hand. With this weapon he will be ready to fire rays that travel at the speed of light when the train and its would-be saboteur pass beneath him. The relativity-crossed Agent X does not know it yet, but he is about to pay dearly for his wicked ways.

15 As Agent X rockets over the roofs of the Relativity Express, he is the target of both guards. At the instant the two guards come abreast they fire their laser pistols. The gun on the train is moving, the other gun is stationary on the overpass, but—because the speed of light is always the same regardless of the motion of its source—both laser beams reach their quarry simultaneously. The hapless X, receiving his harshest lesson in relativity, is struck not once but twice.

16 The great nuclear power plant is saved as the Relativity Express grinds to an unscheduled stop. After frantically searching the clock, a guard finds the bomb, disarms it, and holds it aloft jubilantly. "The bomb," he exults, "was set for 1:30 but it is only 1:00 on the Relativity Express now, for the train clock has been running slow relative to the ground. Thank Heavens for relativity; it has bested the evil Agent X." Meanwhile the benighted agent is stretched out on the ground behind him.

 Foiled by relativity, the once-smug Agent X is hauled away. The train, stopped, measures exactly the distance between two telephone poles; at full speed it was compressed to a fraction of that space (page 160); its wheels are round once more. A guard says pityingly, "Don't take it so hard, X. There's solace for you. Since everything on the train happened slowly relative to the ground, you're actually half an hour younger than you would have been if you hadn't ridden the Express." And so, younger but wiser, X is dragged off to jail.

8
The Arrow of Time

Theoretically, in the situations covered by Newton's laws, forward time and backward time are indistinguishable. For instance, when a film sequence of a tennis match is reversed *(right)*, it is not apparent to a viewer that time is running backward.

TIME, which is everywhere yet nowhere, which is intuitively obvious yet logically indefinable, which is as straightforward as an alarm clock yet as paradoxical as relativity, has inevitably fascinated speculative minds in almost every age. For centuries, philosophers and scientists have chewed over time's puzzling properties and tried to determine what it is and where (if anywhere) it is going.

Their speculations have centered on three fundamental questions: Is time real or unreal? Does it move in one direction only or is it reversible? Does it have a beginning or an end or is it infinite? None of these questions has yet been answered to everyone's satisfaction, and some of them may never be answered to anyone's satisfaction. Yet the mere asking of them stretches the mind, and the search for answers, though it may prove unsuccessful, can still reveal much about both time and the universe we inhabit.

The first question, "Is time real?", can be quickly disposed of. For the scientist, the question is meaningless, in the sense that it cannot be answered by scientific methods, and also trivial, meaning that it is not worth answering. To ask whether time is real is to ask whether the changes—biological, astronomical or atomic—by which we measure time, and which we measure by it, are real. And the scientist, whose business is precisely the understanding and explanation of such changes, is compelled to assume that they—and therefore time— are real. If the changes are not real, or at least real enough to perform experiments with, he is in the wrong line of work.

Beyond this, all of us here on earth—scientists, philosophers, doctors, lawyers and candlestickmakers—live in a world in which time and changes, real or not, must be coped with. A friend of the Russian philosopher Nikolai Berdyaev recounts how the old man would "plead passionately for the insignificance and unreality of time, and then suddenly stop and look at his watch with genuine anxiety at the thought that he was two minutes late for taking his medicine!" A philosopher today who jaywalks across a superhighway, believing that the motion of the speeding cars is an illusion, will not survive very long to philosophize on the matter.

To be sure, most of us, when feeling too beset by the ceaseless changes of existence, have at some point daydreamed of a place where time and change were not—of Tennyson's Lotus Land "where it was always afternoon." With the Irish poet Yeats we have yearned for the Land of Heart's Desire,

> *Where beauty has no ebb, decay no flood,*
> *But joy is wisdom, time an endless song.*

But though in our dreams time may have a stop, it starts up again in the moment of our waking.

A no less familiar chord of longing is struck by the second fundamental question, "Can time be reversed?" Who among all of us has not at some time in his life mused along the lines of the Victorian doggerel,

Backward, turn backward, oh time in your flight;
Make me a child again, just for tonight,

or wished with Shakespeare's Richard II to "call back yesterday, bid time return"? To turn back time, to be able to undo our mistakes and relive our ecstasies—that would be a secret worth having! For such a magical elixir of youth, the alchemists toiled long years over their alembics, and Faust was prepared to sell his very soul.

Common sense born of experience, however, teaches us that a backward-moving time is as fantastical as an unmoving time. The kettle boils only after we have set it on the stove, never before. The apple ripens only after the apple blossom has fallen. All of us grow older with time, not younger.

Proving that time cannot run backward

Yet for science, which has learned the hard way to distrust "what everybody knows," common sense is not enough. One must prove that common sense does not deceive us and make us draw conclusions which are merely the result of the limitations of human experience, like the absolute time which Einstein disposed of. And in fact if we set out to prove that time cannot run backward, the job turns out to be rather harder than we might expect.

When we take a hard look at the basic laws governing physical change, we find that most of them seem to operate equally well in either direction. Change, which is the measure of time, can operate either "forward" or "backward." Among the best examples are the motions of the planets—which were the very first motions to be reduced to mathematical law. If some cosmic movie camera were to film the planets for a year or two, the film would be consistent with the laws of celestial mechanics whether it were run through the projection machine forward or backward. If we did not know to begin with that the planets move around the sun in a particular direction, we would have no way of knowing which version of the film was the "true" one. Again, we *know* that the sun rises in the east—yet there is no reason in physics why the earth might not rotate in the opposite direction, putting sunset in the east and dawn in the west.

Equally reversible in time are the laws of electromagnetism. If a direct current of electricity passes through a coil of wire, the coil will become a magnet. If the direction of the current is reversed, exactly the same thing will happen, except that the north and south poles of the magnet will change places. Neither arrangement is any more "lawful" than the other. As explained in Chapter 5, an atom can emit electromagnetic radiation of a particular frequency—or absorb the same radiation. If the process of emission could be filmed, the same film run backward in time would seem to show absorption, and we could not prove that it did not. All these basic physical laws are, in the technical phrase, symmetrical with respect to time. So far as they are concerned, time can run forward

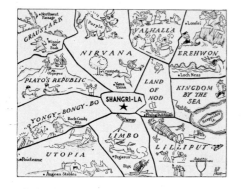

TIMELESS LANDS were satirized in a 1942 newspaper cartoon after President Franklin D. Roosevelt said World War II bombers that hit Tokyo were based in Shangri-La. A fictional place of eternal youth, conceived by James Hilton for his novel *Lost Horizon*, Shangri-La is shown surrounded by other places where time stands still, such as Limbo, home of souls cut off from time, and Valhalla, where Norse warriors live on to glory in battle after battle.

or backward; time's direction makes no difference. If all these elementary processes involving the interactions of waves and particles with each other are symmetrical with respect to time, why not the large-scale, everyday processes on which our commonsense notion of time depends? After all, the large-scale processes are nothing but myriads of microscopic processes happening together.

The symmetrical laws suggest that time could run backward, not that it must in reality do so. Does it? A more definite answer comes from a very common physical process which, as it turns out, is not symmetrical: the transfer of heat.

If we place an ice cube in a glass of water, the ice will absorb heat from the water and glass and will melt: the water and the glass, by giving up part of their heat, will both grow colder. And if a film of this process were run backward, we would notice it immediately. A film in which part of the water in a glass froze into an ice cube, while the rest of the water got warmer, would be instantly detectable as a fake. Heat, that is to say, will flow only from a warmer body to a cooler body, never the reverse.

This and similar observations were generalized more than a century ago by Lord Kelvin (the same man who disputed Darwin's estimate of the earth's age) and the German physicist, Rudolf Julius Emmanuel Clausius, into the second law of thermodynamics. It is this law which, most scientists believe, expresses the one-way nature of time.

The laws of disorganization

The law deals with one of the most basic concepts in physics: entropy. Kelvin and Clausius defined entropy in terms of heat transfer, but the concept was later broadened to encompass the transfer of all forms of energy. The broadened law can be formulated in a number of ways. For our purposes, the most useful is that in which entropy is related to a gain or loss of organization. That is, in any collection, or "system" of things, a loss in the organization—an increase in "randomness"—of their arrangement is equivalent to an increase in entropy.

Taking the system of the glass of ice water, we say that when the ice melts the system has gained entropy because the heat energy in it, instead of being more or less concentrated in the water, is now distributed rather evenly throughout the glass. The ice itself has also gained entropy, because the water molecules that compose it are less organized, when they have melted into the liquid form, than when they were locked into the regular structure of the ice cube. The original water, on the other hand, has lost entropy; its molecules, having given up some of their heat energy to the melting ice, move around less freely and vigorously.

The second law of thermodynamics, then, says that in any system, some constituents (in this case the ice) may gain entropy with time, others (the water) may lose entropy. But the entropy of the whole system (the ice and water together), if it changes at all, *must increase with time.* This one-way direction for entropy carries with it a one-way direction for time, since entropy changes will reveal to us when "the film is running backwards."

When a child grows into an adult, he has lost entropy, because the very complex organization of atoms that is his body has grown larger. But he achieves this entropy loss only by breaking down—"disorganizing"—the chemical structures of the food he eats. And the entropy gained in the breakdown of his food exceeds the entropy lost in the buildup of his body. (Much of the food energy, for example, is transformed not into tissue but into body heat, which eventually dissipates into the atmosphere.) Taking the child and his food as a whole, therefore, entropy has increased.

What is true of children is also true of clocks. In the most perfectly adjusted chronometer, the input of energy exceeds the mechanical output; part of the energy input is transformed into heat by friction of the gears and by the bending back and forth of the balance spring; the heat energy, dissipated into the atmosphere, represents a gain of entropy. In an atomic clock, entropy is increased when part of the electrical input is transformed into heat by electrical resistance.

Thermodynamic gambling

Entropy and other characteristics of energy are often described in gambling terms. The first law of thermodynamics, which declares that energy cannot be created or destroyed, has been informally phrased as "You can't win!" The second law, then, amounts to saying, "You can't even break even!" Entropy, in this sense, is the bookmaker's vigorish, the roulette wheel's house percentage, the irreversible toll which time extracts from the operations of change.

The parallel with gambling is not accidental. A tossed coin, for example, will come up heads or tails, and no one can predict which it will be on any particular toss. But we can predict with great certainty that after a large number of tosses, about one half of the coins will come up heads. And the greater the number of tosses the more accurate our prediction will become. It is along these lines, from the laws of probability that govern the fall of dice and the appearance of numbers at roulette, that we may hope to find an explanation of the second law of thermodynamics. Then we will see why it is not a paradox that the elementary laws of interaction can be symmetrical with respect to time and yet lead to a full-scale world in which time, like an arrow, moves in only one direction.

To understand how the laws of probability influence the laws of time, let us, as we did with relativity, conduct a thought experiment. Imagine a perfectly smooth, frictionless billiard table with no pockets but with cushions that are perfectly elastic. Place on it a billiard ball that is also perfectly elastic. With such a setup, the ball, once it has been set into motion, will keep moving forever, bouncing back and forth across the table from one cushion to another. The motions of this single ball are symmetrical with respect to time; a motion picture film of the ball and table would be equally plausible whether the film was run forward or backward. The same thing would be true, of course, if we start with two balls; they will sometimes carom off each other instead of off the cushions, yet the motions of the two balls will still be time-symmetrical.

But suppose we equip our table, not with one or two balls but with 40. Let us draw a line across the middle of the table, dividing it into halves, and place all 40 balls on the left-hand half. The table is now analogous to a state of low entropy. Its two halves can be thought of as two objects, one of them hot (the side with the balls), the other at the cold of absolute zero (the empty side).

A scientific game of billiards

Let us now start the 40 balls rolling, in random directions. As they carom off one another and off the cushions, some will remain on the left-hand half of the table but others will cross the line into the right-hand half. In the course of time, the distribution of balls between the two halves will inevitably become more uniform—that is, the entropy of the system will increase. Eventually the table will approach a state of maximum entropy, with equal (or nearly equal) numbers of balls on either side of the line.

The reason lies in the laws of chance. So long as there are balls only on the left half of the table, any ball that "changes sides" must inevitably move to the right half. There are no balls that could make the reverse journey. Moreover, so long as there are *more* balls on the left side of the line than on the right, it is more probable that a ball will travel from left to right, further evening the distribution, rather than in the reverse direction, further unbalancing it. But once each half has an equal number of balls, it is just as likely that a ball will roll in one direction as in the other. Thereafter, balls moving from left to right will, on the average, be balanced out by those moving in the opposite direction.

Thus simple probability ensures that the table, starting with a totally unbalanced distribution (low entropy) will sooner or later reach a balanced distribution (maximum entropy). And this process is *not* time-symmetrical, because the reverse process, in which all the balls end up on one side of the line, will never occur.

Never? Well, hardly ever. Paradoxically, the laws of probability also tell us that if enough time elapses we will occasionally find all the balls on one side of the line for a brief moment. At such a moment the system is indeed back in its original low-entropy state. The odds that this strange result will appear can even be calculated for different numbers of balls.

With one ball, the odds on a state of low entropy are one in one—a dead certainty. The ball, except for the instant that it crosses the line, *must* be on one side or the other. With two balls, the odds drop to one in two; with three balls, to one in four. With 10 balls, the odds drop to one in 512, and with 40 balls, to something like one in $10^{12}=10$ followed by 11 zeros!

A Japanese-American physicist, Satosi Watanabe, actually carried

FATHER OF ONE-WAY TIME, Rudolph Clausius *(above)* was a lecturer in physics at the University of Berlin *(below)* when he developed the concept of entropy, which demonstrates that the flow of heat energy always proceeds in one direction. Although Clausius himself did not apply this principle to time, other physicists did: Since heat's energy flow and the flow of time are known to occur together, time also should follow a one-way course.

out an experiment of this sort. He did not, of course, use billiard balls, but a computer, programmed by a system known appropriately as the Monte Carlo method. Watanabe's computations, translated into billiard-table terms, assumed that each ball had, on the average, one collision every second, and that after each collision it had one chance in 50 of crossing the dividing line between the two halves of the table. Starting with 10 balls on one side, the system could reach maximum entropy (i.e., a 5-5 distribution) in, say, 55 seconds. In only 70 seconds more, however, the system would be back to minimum entropy, with all the balls on one side. With 100 balls, maximum entropy would be reached almost as quickly — in less than 70 seconds. But the entropy would then show no significant drop in several hundred seconds. In fact this particular system should return to minimum entropy, on the average, only once in 15,000 billion billion years. This is about 100,000 times the half-life of lead 204, whose decay was cited in Chapter 6 as the longest natural process.

Thus while it is possible that entropy (and therefore, in a certain sense, time) will reverse itself simply by chance, it is — to put it mildly — highly unlikely, so long as the system involved contains more than a few elements. It is possible that at some moment the moving molecules of the air will so arrange themselves that they will strike our bodies on one side but not the other — thus knocking us sideways — but it is highly unlikely. If on the other hand, we consider not a man but a grain of dust, which at any moment is not being struck by countless air molecules but only by a few, we find that it *is* being knocked around and almost constantly at that. This phenomenon, called Brownian movement, was discovered by the Scottish botanist Robert Brown in 1827. When he examined a drop of water containing pollen grains under a microscope, he found that the grains were jerking back and forth in continual irregular motion. The motion was not explained until 1905, when Einstein and another physicist showed that it must be caused by more molecules hitting one side of the grain than the other.

The problem of reversing the universe

Thus, for tiny portions of the universe, containing very few objects — such as 10 billiard balls or one pollen grain — temporary reversals of entropy are probable, even routine. But for a system of any size and complexity (the case of 100 billiard balls), the probability becomes so small that it can be considered nonexistent. And for the largest of all systems, the universe in which we live, the chance that entropy will reverse is correspondingly small.

Entropy, which seemed to give direction to time's arrow, offers merely a preferred direction. This is the only direction that human experience has ever recorded, the only one that can yet be tested by experiment and the only one that is at all likely, but it is not the only one that is conceivable. The faint possibility that time could be reversed has made time-travel a favorite subject of speculation among writers and readers of science fiction. One kind of time machine simply turns everything backwards, like a movie film running the wrong way. The probability of such a large-scale time reversal is, for the reasons explained above, unimaginably small. But even assuming that it did occur, the question remains whether this kind of time-travel would subject the traveler to true time reversal or merely to a reversal of the order of events. If, like the reader of the story, he were merely a bystander observing the operation of the machine, he would see very strange phenomena, but his own time would run as usual.

There is, to be sure, another speculative version of time-travel in which the past and the future both exist at the same time as the present, so that a time machine could carry a traveler back and forth between them. It is true that we see faraway stars as they were millions of years ago and we ought to be able to tap this reliable source of information. A rather similar argument applies to projection into the future. There are those who believe that the laws of nature are so rigid that every future event is irrevocably fixed. Thus the future could be perceived by a super-mind or a super-machine.

The end of time

When we turn from time reversal to our third question, "Does time have a beginning and an end?" the answers become considerably vaguer. Definite statements would require that we first settle some other fascinating but elusive questions, such as "What do you mean by a beginning or an end?" "Is the universe finite or infinite?" and, for that matter, "What do you mean by the universe?" Yet strangely enough, partial answers can be suggested. For convenience, it is best to begin with the "end" of time rather than the "beginning."

Throughout this book time has been defined in terms of change. If we ask, therefore, whether time has an end, we are in effect asking if change has an end. Considering the link between change and entropy, it seems as if it ought to. As a matter of fact, the Austrian physicist Ludwig Boltzmann long ago visualized the end of the universe (and therefore the end of change and time) as the attainment of maximum entropy by the universe as a whole. At some undefined date in the future, according to this conjecture, nothing will be hotter or colder than anything else. The slowest radioactive elements will have decayed into stability. The stars will have radiated away their furious energy, warming the frigid dust of interstellar space to a fraction of a fraction of a degree. Earth and its sister planets, their rotation slowed by friction with cosmic dust and

WHY TIME GOES FORWARD is shown by an experiment which illustrates the relation of time flow to changes in the universe's energy distribution. The experiment requires three jars, 40 numbered slips of paper and 40 numbered balls representing energy. At the start all the slips are in jar 1, all the balls in jar 2 and nothing in jar 3. A slip is drawn at random *(extreme left)* and the ball having the same number is transferred to jar 3 *(center)*. The paper is replaced and the process is repeated; each time a ball is moved either from jar 2 to jar 3 or vice versa. After about 25 moves, the jars will hold nearly equal quantities of balls *(right)*. This results from the mathematical laws of chance, which govern (1) the numbers that are drawn, (2) the direction of energy flow and (3) —since energy flow and time flow proceed together—the direction of time.

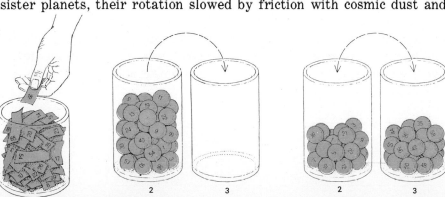

gases, will have fallen out of orbit into the sun. Man himself will be long extinct. In this burned-out universe, there will be no change by which time can be observed or measured. In some abstract, metaphysical sense, time may still exist, but scientifically speaking, it will have ended.

If this is to be the end of the universe and of time, it is possible to guess when this end will come. The slowest known process involving energy transfer (and therefore entropy) is the decay of lead 204, which, as already noted, has a half-life of 140 million billion years. There are about four million billion tons (3.6 million billion t) of lead 204 in the earth, figuring from the mass of the earth and the prevalence of the isotope in rocks that have been analyzed. In truth, the actual amount does not matter much, even if we attempted to consider all the lead 204 in the entire universe. No matter how much there is, half of it will be gone in 140 million billion years. And if this halving process is repeated about 1,000 times, the lead will be gone to the last atom. Multiplying the half-life by 1,000, then, gives a very rough "outside" date for the end of time: A.D. $10^{20} = 100$ billion billion years from now.

This assumes, of course, that nothing will happen in the meantime to transform the lead into something else. But it is quite conceivable that something of the sort *will* happen. To see what that something might be, we must look backward toward the "other end" of time—the beginning, rather than the end, of the universe.

Starting off with a bang

If the universe is changing, how and when did these changes start? The obvious answer is with the "big bang," which, as Chapter 6 outlined, has been deduced as the initial motive power behind the universe's expansion. Scientists like this theory because it accounts quite well for the known facts about the structure and changes in the universe. By its reasoning, the big bang would be the beginning of time along with everything else—the Creation, in fact.

But there is an old story of a little boy who naively asked: "If God made the world, who made God?" Likewise, there are scientists, and not naive ones either, who insist on asking, "What made the big bang?" Their answer provides a picture of the universe which is, to some people, the most esthetically satisfying of all.

The scientists who look beyond the big bang concede that the universe is expanding—at present. But eventually, they suggest, the mutual gravitation of the receding galaxies, feeble though it certainly is, will slow the expansion, halt it and, finally, reverse it. Perhaps 40 billion years after the big bang, the universe will begin to contract. The result, after another 40 billion years, will be—another big bang, *which will start the cycle all over again.*

Thus the bang which began "our" universe was also the bang which ended a previous one; the end of our universe will also be the beginning of another. The universe, like the fabled phoenix, periodically perishes in fire and is reborn from its own ashes. The end of time is also the beginning, and the universe is its own ultimate clock, ticking off its majestic cycle of expansion and contraction in 80-billion-year periods. The existence of such a universal cycle is as yet supported only by extremely tenuous evidence. It includes, among other things, estimates of the rate at which the most distant known galaxies, at the very edge of the observable universe, are receding from the earth. These measurements, some astronomers believe, indicate that the expansion of the universe is in fact slowing down. But their interpretation is blurred by the fact that at those remote distances the astronomers cannot clearly distinguish the effects of space, time and motion. The motion of these far galaxies is, to be sure, inferred from their red shift—but their distance is partly inferred from their motion, on the theory that the greater the red shift, the greater the distance. Finally, their place in the universe's time scale is also inferred from their distance. We see galaxies at a distance of five billion light-years, not as they are now but as they were five billion years in the past—and if the estimate of distance is off, so is the estimate of time.

The universe before ours

Even less can anyone guess how many times this rejuvenating cycle has occurred or will occur—assuming it exists at all. Some scientists, speculating on the nuclear processes which are thought to have formed the elements, and noting the seemingly inexplicable abundance of some heavy elements, argue that our universe must have been formed from the debris of a previous one—i.e., that there must have been at least one cycle before the present one. But the question is far from settled and perhaps will never be settled. So far as the cyclical universe is concerned, it is possible to say only, with James Hutton, "we find no vestige of a beginning, no prospect of an end."

But suppose the cyclical theory is wrong? Suppose the universe ends, not with a revivifying bang but with the whimper of Boltzmann's maximum-entropy state. Will time be ended for good? Probability suggests that it may not be. Just as all the billiard balls will sooner or later find themselves back on one side of the table, so even a dead universe will, simply through the laws of chance, once more reach a state of low entropy.

How long this chance rejuvenation might take is beyond calculation or even imagination. If 100 billiard balls on a table take more than 10,000 billion billion years to return to low entropy, how long will all the billiard balls, molecules and atoms in the universe take? No one can guess.

But after that night of inconceivable duration, in which time and change have stopped, they may yet begin once more, and once more may beget intelligent beings who, like us, speculate on the paradoxes and mysteries of time. And if this happens, it is not unlikely that one of these creatures will, in time, echo in its own words the thought of the philosopher-mathematician Alfred North Whitehead: "It is impossible to meditate on time and the mystery of the creative passage of nature without an overwhelming emotion at the limitations of human intelligence."

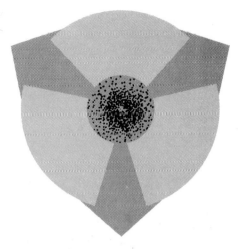

WHEN TIME BEGAN is explained by the theory diagramed above. It assumes that the present universe started with a huge explosion *(center)* which caused matter to expand outward *(dark circle)*; the observed rate of expansion sets the origin of the universe about 16 billion years ago. This theory holds that expansion will continue indefinitely, as indicated by large arrows. A second theory, illustrated below, maintains that there was a time span even before the existence of our universe. It postulates repeated cycles of expansion and contraction *(double arrows)*, each cycle lasting about 80 billion years.

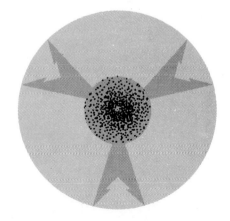

Man versus Clock: the Unequal Struggle

Ticked off on a clock, and crossed out on a calendar, time moves inexorably forward. And yet human nature cannot yield to this fact. During his brief passage on earth each man wages a ceaseless though silent struggle to defeat time—to manipulate it to his private purposes by moving it back or hastening it forward; to wrench from its grasp an instant that may be relived and savored time and again through recollection; to stave off—if but for a moment, a day, a month, a year—"Time's winged chariot hurrying near." The youngster reading of heroes long gone is transported in his mind to another age and sees himself storming the gates of Carthage or Troy or rescuing a fair-haired damsel from a dragon's jaws; the matron casts off the years—at least to her own satisfaction—with every trip to the beauty parlor; the mystic consulting the stars believes he can foretell the future. Only death ends the individual's struggle against the passing hours. And death itself is not yet the end. For nearly everyone seeks a measure of immortality—physical, through the passing of genes to his children; historical, through achievements left for future generations; or spiritual, through religion's promise of "life everlasting"—that will forever break the grip of time.

PRESERVING A HAPPY MOMENT
Three generations of a Polish family pose before a backdrop depicting the city of Czestochowa, where they have spent the day. For each member of this group portrait, seeing the picture will roll time back to younger days, reviving the pleasures of a day in town—just as promised by the backdrop inscription. "Souvenir from the Jasna Góra Cloisters in Czestochowa."

Bits and Pieces of Recorded Time

In a variety of ways, man tries to bring the past into the present. Some people paste photographs in albums. Others preserve old ticket stubs, antiques, back issues of old magazines.

Nostalgia is not the sole reason for saving: Mementoes are man's most meaningful link with the past. The fossils at right enable scientists to study the life of almost a billion years ago. In the National Archives Building in Washington, D.C., historians can examine 900,000 cubic feet (25,485 m³) of government records dating to the nation's beginning and including such documents as the Declaration of Independence and the Constitution. Among the 60 million objects in Washington's Smithsonian Institution are General Sheridan's horse and John Glenn's space capsule.

Nothing is too large or too trivial to be preserved. In the London Museum, British statesman William Pitt's saucepan is on display. In the U.S. whole communities—Colonial Williamsburg, Virginia; Mystic Seaport, Connecticut; Old Sturbridge Village, Massachusetts—have been restored to make the past live again.

THE COLLECTED PAST
Surrounded by carefully compiled scrapbooks, accounting clerk Ray Smith, an inveterate collector of personal memorabilia, relives his earlier years. Included in his collection are toys from his childhood, picture albums filled with high school photographs, a stack of programs from plays he has attended, old magazines, party favors, and high school and college pennants.

FOSSILS AND ARTIFACTS ON FILE
An open drawer at the American Museum of Natural History in New York contains the fossilized remains of marine animals ranging in age from 50 million to about 700 million years. Beyond the drawer and in other rooms of the museum millions of carefully catalogued and filed artifacts and relics date from before the dawn of civilization to the contemporary era.

THE MEETING OF TWO WORLDS

An Amish couple *(right)* in Lancaster, Pennsylvania, dressed in the simple black garb their faith decrees, head for home after a shopping trip to town. Forbidden by their religion to possess or even ride in automobiles, the Amish travel by horse-drawn buggy instead *(below)*.

A Way of Life Cut Off from Time

Now and again nearly everyone has wished he could turn back the clock to a simpler and more pleasant age. Some people immerse themselves in earlier times by reading or studying history. But few have succeeded so well in disassociating themselves from the flow of time and creating their own world as the Amish sects of Pennsylvania, Ohio and Virginia.

Responding to a religious injunction that "a separation shall be made from the wickedness which the devil planted in the world," the strict Old Order Amish live completely apart from the religious, social and political activities around them, refusing to vote, swear oaths in court or bear arms. They do without factory-made gadgets and modern conveniences such as electricity, TV, modern dress, buttons, mirrors, stuffed furniture and pictures on the wall. Some contact with the outer world, on the highways and in stores, is unavoidable, but the Amish stubbornly resist every pressure to make them conform to the times of the society around them.

The Costly Effort to Turn Back Time

In 1513, Ponce de León, ripe with his 53 years, hobbled into the Florida jungle in search of the elusive fountain of youth. The impulse that sent him on his futile quest finds many echoes in the modern world. Old age is so abhorred that Americans try to deny its existence by calling the elderly "senior citizens." The aging wish away the advancing years with such hopeful phrases as "you're only as old as you feel," and women are sometimes granted the legal privilege of concealing their age. In their determination to turn back time and be more chic, American women spend more than $5 billion annually in the nation's beauty salons. The more privileged among them repair to posh resorts where for fees of close to $2,000 per week they spend their hours in splendid misery trying to obscure time's ravages by being poached in steam, prodded by lady drill instructors, rubbed with turtle oil, cooked in linens, bathed in milk and fed dry toast off exquisite china.

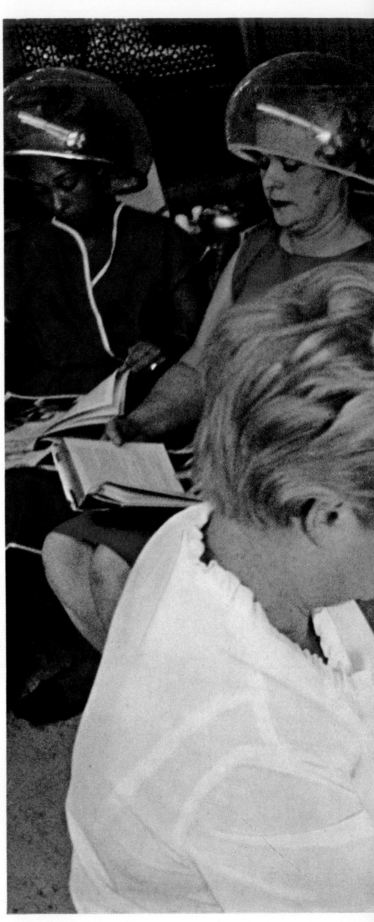

PEELING OFF THE YEARS
Stretched out on a health club floor *(above)*, seated beneath hair dryers *(opposite)* or engaged in crash diets, American women spend time, money and effort trying to stay young.

182

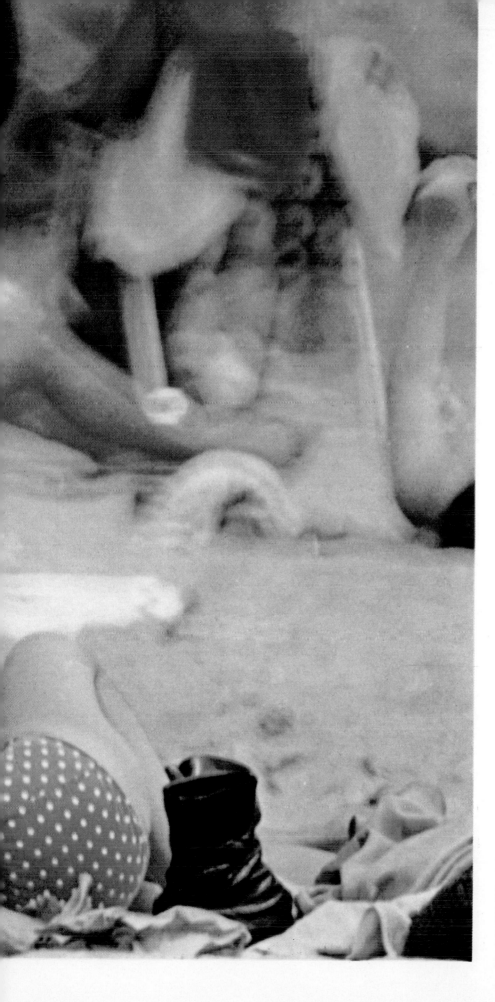

When Time Stands Still

Beginning at least as far back as the Roman poet Ovid—"One would think," he wrote, "that time stood still, so slowly does it move"—novelists, poets and songwriters have celebrated the "timeless moment." Almost everyone recalls the rare event so imbued with deep feeling that it carried him into a private world of joy or sorrow that seemed without time or space. The event may be of national importance: a critical election for an important political office, a tense emergency. Or it could be purely personal: the birth of a child, the death of a relative or even an evening walk that brings sudden awareness of the loveliness of nature. And sometimes a kiss can stop the clock, granting what George Bernard Shaw called "eternity in a single moment."

A TIMELESS EMBRACE
Locked in a close embrace, a boy and girl momentarily set aside time and reality. Around them swirls a Fourth of July crowd at one of the nation's busiest beaches, New York's Coney Island, but for an enraptured instant the couple are blind to everything but themselves.

NAPOLEON'S MISFORTUNE TELLER
A modern edition of a book supposedly used by Napoleon features his palm on its cover. The long vertical line in the center of the palm connotes misfortune; the ladder-like lines on the little finger warn of a powerful enemy, England.

A Murky Look into the Future

The impulse to peer into the future and see what fate decrees is probably as old as mankind and it continues to exert a powerful hold on millions. On February 3, 1962, thousands in India fled their homes in panic, convinced by astrologers that the world was about to meet its doom.

Many notorious and famous people have also taken astrology seriously. So astute a financier as J.P. Morgan often consulted an astrologer; the famous psychiatrist Carl Jung had horoscopes drawn up for his patients. In Nazi Germany Joseph Goebbels and Heinrich Himmler were so fascinated by soothsayers that Winston Churchill hired his own astrologer to tell him what his enemy's stargazers were predicting. Holland was once wracked by a crisis of state when it was discovered that Queen Juliana had been receiving advice from a mystic, and in Hollywood one fashionable seer counts many movie stars among his devotees; his column, in more than 125 newspapers, is a guide into tomorrow for thousands of readers.

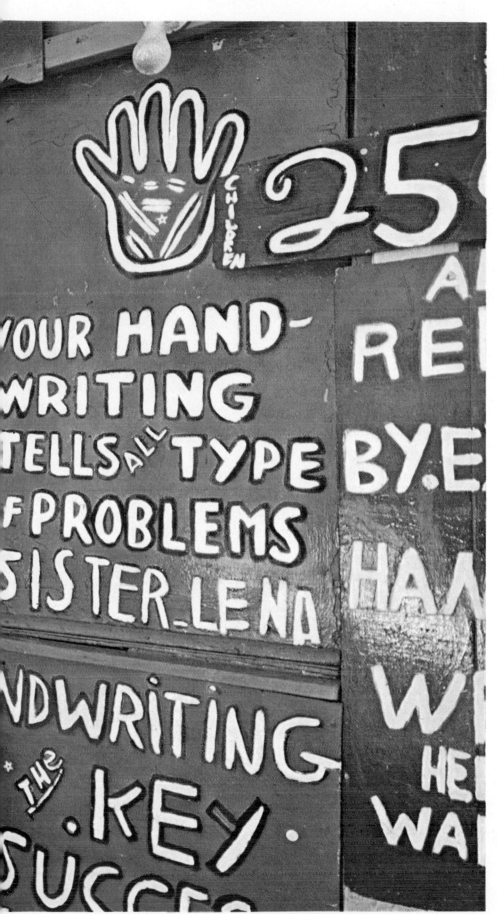

A LOOK AT WHAT LIES AHEAD
Sister Lena, a fortune-teller at the Coney
Island, New York, amusement park, examines a
young man's palm *(above)*, informs him of his
fate and then, for an additional fee, advises him
how he may alter the future *(left)*. Though they
are often derided, fortune-tellers flourish. In
New York City alone, according to police estimates, there are more than 400 of them.

Monuments to Immortality

Soaring over the dismal Watts section of Los Angeles are seven spidery constructions, each looking like a confectioner's version of the Eiffel Tower. These strange and twisted shapes are a monument, built over a 33-year span by an immigrant Italian laborer, Simon Rodia, who later remarked, "A man has to be good good or bad bad to be remembered."

Rodia's private urge for immortality through remembrance echoes the desire of nearly all men, obscure or famous. Ramses II, Pharaoh of Egypt, pursued the same goal with relentless zeal, raising obelisks, temples and statues as monuments to his own magnificence. Ramses went so far as to efface the names of earlier monarchs from their own monuments, replacing them with his own. The key to his behavior lay in an ancient Egyptian proverb—"to speak of the dead is to bring him to life."

RODIA'S ARTISTIC TRIUMPH
The steel tracery of Simon Rodia's towers, built between 1921 and 1954, rises above Los Angeles. The tallest *(foreground)* soars 90 feet (27 m). Some years ago the towers were about to be torn down but art lovers protested, and the towers were declared a "cultural monument."

A JUNK-MADE MOSAIC
The base of a Rodia tower, covered with a mosaic of seashells, tile chips and broken glass, sparkles in the sun. The fanciful garden includes several birdbaths, a gazebo and a model ship, as well as the towers. All are richly decorated with mosaics formed from the refuse Rodia collected over the years from junkheaps. One art critic has estimated that the construction materials included approximately "7,000 sacks of cement, 75 thousand seashells, hundreds of broken dishes, thousands of pieces of tile and several truckloads of broken bottles."

The Promise of Eternal Life

"Whosoever believeth in Him should not perish, but shall have everlasting life," says the New Testament. This promise of immortality, of the ultimate triumph of the individual over time, is at the core of many religions, though it is interpreted in a variety of ways. To some the promise means a physical rebirth in God's Kingdom, a land above the skies, a better world, free of the injustices and frustrations of this earthly existence. Others see the hereafter as a more subtle state in which the individual soul merges with the cosmos. In Hinduism men are thought to be reincarnated here on earth in animal or human form—an evil man might return as a snake, a good man as a maharajah.

But however the promise of an afterlife is viewed it always expresses man's refusal to accept death as the total oblivion of self—as "deserts of vast eternity." It is this belief that, at the close of man's own personal, physical span, gives him his last and best opportunity to conquer time.

IN PRAYERFUL SUPPLICATION
Protestant congregants kneel in prayer at a Christmas Eve service at New York City's Episcopal Cathedral of St. John the Divine. Central to the faith of these worshipers is the belief that God in His mercy rewards the worthy through the salvation of their immortal souls.

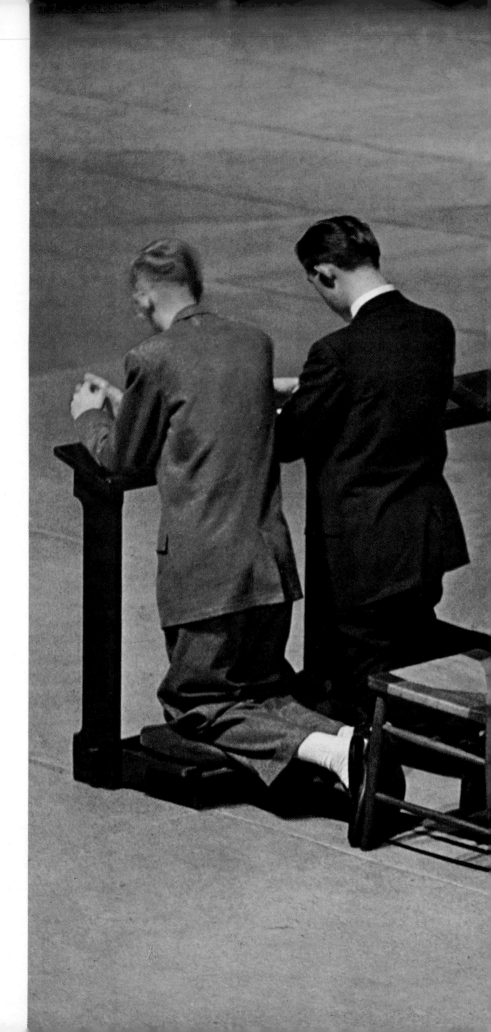

Perfecting the Clock

ACCURACY IN TIMEKEEPING has improved steadily, as shown below, since the first mechanical clocks were made in the 14th century. Early clocks, like the Dover Castle *(lower left)*, were regulated by a weighted bar called a foliot, which pivoted back and forth to move a single hand. Although foliot mechanisms became increasingly precise over the next three centuries, clocks still varied by several minutes a day until pendulum clocks came into general use in 1656. Then, for the first time, clocks became accurate enough to record minutes as well as hours.

Over the next 265 years clockmakers developed better escapements for regulating the pendulum, and then began to improve the pendulum itself. In 1721 George Graham was the first to compensate for the fact that temperature changes cause steel pendulums to vary in speed. His clock *(lower right)* had a tem-perature-independent mercury-vial pendulum, which varied by only one second a day. Accuracy continued to improve as pendulums were superseded by quartz crystals which vibrated so precisely that they made possible a clock accurate to a few hundred thousandths of a second per day.

Even more precise clocks came into being in the 1950s, when atomic oscillations were used to regulate the vibrations of a quartz crystal. Atomic clocks paced by vibrating cesium atoms were perfected until they were precise to several billionths of a second per day, or about one second every 350,000 years. For time periods shorter than a day, as the graph shows, hydrogen maser clocks were even more exact than the other types. However, for measuring longer periods of time the cesium clock remains the most accurate timepiece that man has devised.

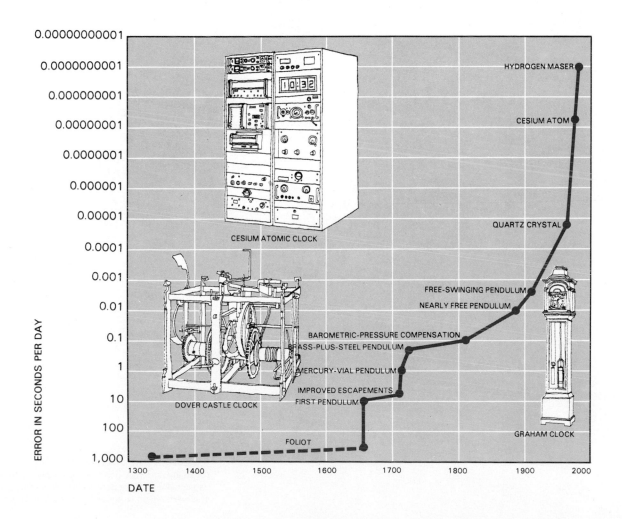

ERROR IN SECONDS PER DAY

DATE

CESIUM ATOMIC CLOCK

DOVER CASTLE CLOCK

FOLIOT

FIRST PENDULUM
IMPROVED ESCAPEMENTS
MERCURY-VIAL PENDULUM
BRASS-PLUS-STEEL PENDULUM
BAROMETRIC-PRESSURE COMPENSATION
NEARLY FREE PENDULUM
FREE-SWINGING PENDULUM
QUARTZ CRYSTAL
CESIUM ATOM
HYDROGEN MASER

GRAHAM CLOCK

SECOND HAND

MINUTE HAND

HOUR HAND

BARREL ASSEMBLY

CROWN

MAINSPRING

ESCAPE WHEEL

A MECHANICAL WATCH
As many as 160 parts fill the casing of a traditional watch. It operates on a few billionths of a watt, which is supplied by the mainspring.

The Workings of Two Watches

Like the clock, the watch has become more streamlined and accurate with the years. Whether it is of the mechanical hands-display type *(left)* or the electronic digital-display type *(below)*, it is a triumph of minute technology, one of mankind's smallest, sturdiest, most precise machines.

The driving power for the mechanical watch is supplied by the mainspring *(in blue at center left)*, which is tightened by a few twists of the windup knob, or crown *(far left)*. The mainspring is mounted on a large gear called the barrel assembly. As the barrel assembly unwinds it turns the shaft of the minute hand and also a series of gears that propel the hour and second hands.

The motion of the barrel assembly is regulated by an ingenious device called an escape wheel. On the escape wheel are teeth, which are locked and unlocked by the rocking of a tiny lever—the pallet. The action of the pallet is paced by oscillations of a tiny hairspring mounted on a balance wheel. When the balance wheel is correctly adjusted, the hairspring oscillates exactly 300 times a minute, no matter how much pressure the pallet applies, thus doling out motion to the gears in proper amounts.

A battery-powered electronic watch *(below)* is paced not by a hairspring but by a tiny quartz crystal, which oscillates precisely 32,768 times a second. The crystal is connected to an integrated circuit—consisting of many minute electrical switches very much like those in a computer—which simply counts oscillations. When the circuit has counted a minute's worth of oscillations, an electrical signal passes to the digital display, an arrangement of lights which then changes to indicate the new time.

The quartz crystal has given the watch new standards of precision. A well-made mechanical watch may gain or lose only a few seconds a day, but a good electronic watch can be accurate to within a minute a year.

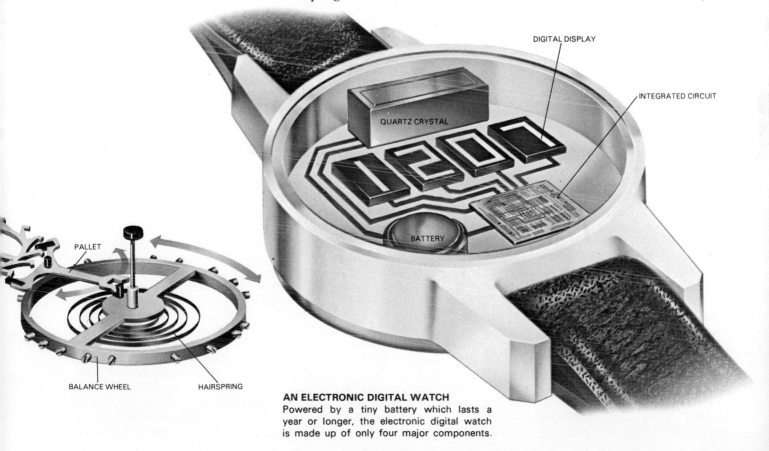

PALLET

BALANCE WHEEL   HAIRSPRING

DIGITAL DISPLAY

INTEGRATED CIRCUIT

QUARTZ CRYSTAL

BATTERY

AN ELECTRONIC DIGITAL WATCH
Powered by a tiny battery which lasts a year or longer, the electronic digital watch is made up of only four major components.

FURTHER READING

The Physics of Time

Barnett, Lincoln, *The Universe and Dr. Einstein*. Bantam Books, 1974.
Bondi, Hermann, *Relativity and Common Sense*. Doubleday, 1964. *The Universe at Large*, Doubleday, 1960.
Borel, Emile, *Space and Time*. Dover Publications, 1960.
Gamow, George, *Mr. Tompkins in Wonderland*. Cambridge University Press, 1967. *One, Two, Three ... Infinity*. Viking Press, 1963.
Landau, L.D., and G.B. Rumer, *What is Relativity?* Basic Books, 1961.
Rosenberg, G.D., and S.K. Runcorn, *Growth Rhythms and the History of the Earth's Rotation*. John Wiley & Sons, 1975.
Schlegel, Richard, *Time and the Physical World*. William Gannon, 1967.

Man and Science

Eicher, Don L., *Geologic Time* (2nd edition). Prentice-Hall, 1976.
Hawkins, Gerald S., *Stonehenge Decoded*. Doubleday, 1965.
Kramer, Samuel N., *History Begins at Sumer*. Doubleday, 1959.
Neugebauer, Otto, *The Exact Sciences in Antiquity* (2nd edition). Brown University Press, 1970.
Price, Derek D., *Science Since Babylon*. Yale University Press, 1975.

The Rhythms of Life

Beck, Stanley, *Insect Photoperiodism*. Academic Press, 1968.
Blum, Harold F., *Time's Arrow and Evolution* (3rd edition). Princeton University Press, 1968.
Bünning, Erwin, *The Physiological Clock* (3rd edition). Springer-Verlag, 1973.
Conroy, R.T., and J.N. Mills, *Human Circadian Rhythms*. Longman, 1971.
Fraisse, Paul, *The Psychology of Time*. Greenwood Press, 1976.
Luce, Gay Gaer, *Body Time*. Bantam Books, 1973.
National Academy of Sciences, *Biochronometry*. 1971.
Ward, Ritchie R., *Living Clocks*. Alfred A. Knopf, 1971.

Science and Philosophy of Time

Eiseley, Loren, *The Firmament of Time*. Atheneum, 1960.
Fraser, J.T., and N. Lawrence, eds., *The Study of Time*. Springer-Verlag, 1975.
Moore, Wilbert E., *Man, Time and Society*. R.E. Kreiger, 1963.
Priestley, John B., *Man and Time*. Doubleday, 1964.
Shapley, H., et al., *Time and Its Mysteries*. Macmillan, 1962.
Toulmin, Stephen, and J. Goodfield, *The Discovery of Time*. Harper & Row, 1965.
Whitrow, Gerald, *The Nature of Time*. Holt, Rinehart & Winston, 1973.

The Mechanics of Time

Chamberlain, Joseph, *Time and the Stars*. Doubleday, 1964.
Clutton, Cecil, et al., *Britten's Old Clocks and Watches, and Their Makers* (8th edition). E.P. Dutton, 1973.
Gould, Rupert T., *Marine Chronometer*, Albert Saifer, 1966.
Harrison, Lucia Carolyn, *Sun, Earth, Time and Moon*. Rand McNally, 1960.
Mayall, R. Newton, and Margaret Mayall, *Sundials*. Sky Publishing Corp., 1973.
O'Neil, W.M., *Time and the Calendars*. Sydney University Press, 1975.

ACKNOWLEDGMENTS

The editors of the revised edition of this book are especially indebted to Gary Catlin, Litronix Corporation, Cupertino, California; Sidney Horenstein, Scientific Assistant, Department of Fossil and Living Invertebrates, American Museum of Natural History, New York; Richard Krasnow, Research Fellow in Biology, Harvard University, Cambridge, Mass.; Arthur Winfree, Associate Professor of Biology, Purdue University, West Lafayette, Indiana; and Gernot M.R. Winkler, Director, Time Service Division, U.S. Naval Observatory, Washington, D.C.

Consulting editors for the first edition were Rene Dubos, Emeritus Professor of Pathology, The Rockefeller University, New York; Henry Margenau, Eugene Higgins Professor of Physics and Natural Philosophy Emeritus, Yale University, New Haven, Connecticut; and the late C.P. Snow, novelist and Fellow of Christ's College, Cambridge University, Cambridge, England. The following persons and institutions also provided valuable assistance: Asger Aaboe, Associate Professor of History of Science, Edward Deevey Jr., Professor of Biology, Derek Price, Associate Professor of History of Science, Yale University, New Haven; Henry Albers, Professor of Astronomy, Francis Ranzoni, Professor of Botany, Vassar College, Poughkeepsie; Alexander Alland Jr., Asst. Professor of Anthropology, Maan Madina, Senior Lecturer, William Nethercut, Asst. Professor of Greek and Latin, Robert Novick, Professor of Physics, Wallace Broecker, Professor, George Parks, Electronics Serviceman, David Thurber, Director, Radiocarbon Laboratory, Lamont Geological Observatory, Columbia University, New York; William Balet, James Comley, Consolidated Edison Co., New York; Bryant Bannister, Director, Laboratory of Tree-Ring Research, University of Arizona, Tucson; Elso Barghoorn, Curator, Paleobotanical Collection, Harvard University, Cambridge; Silvio Bedini, Asst. Director, Museum of History and Technology, Smithsonian Institution, Washington, D.C.; Erwin Bünning, Botanical Institute, Tübingen, Germany; Junius Bird, Curator, South American Archeology, Gordon Ekholm, Curator, Mexican Archeology, Kenneth Franklin, Astronomer, Hayden Planetarium, Vincent Manson, Asst. Curator, Mineralogy, American Museum of Natural History, New York; Mary Burns, Film Scanning and Measurement Supervisor, Joseph Fineman, Asst. to the Editor of *The Physical Review*, J.B.H. Kuper, Chairman of Instrumentation and Health Physics, Kwan-Wu Lai, Asst. Physicist, A.W.K. Metzner, Acting Director of *The Physical Review*, Thomas Schumann, Research Associate, Brookhaven National Laboratory, Upton, New York; Phillip Clark, Public Relations, Field Museum of Natural History, Chicago; Harold Edgerton, Professor of Electrical Measurements, M.I.T., Cambridge; Dr. David Ehrenfeld, Archie Carr, Graduate Research Professor, University of Florida, Gainesville; Dan Eichner, Eichner Instrument Company, Clifton, New Jersey; David Finkelstein, Professor of Physics, Yeshiva University, New York; William Fowler, Professor of Physics, Guido Munch, Professor of Astronomy, Maarten Schmidt, Professor of Astronomy, California Institute of Technology, Pasadena; Sheldon Freud, Office of Aviation Medicine, FAA, Washington, D.C.; General Radio Co., Bedford, Mass.; Dr. Joseph Goodgold, Dr. Neil Spielholz, Institute of Rehabilitation Medicine, N.Y.U. Medical Center, New York; Esther Goudsmit, Post-Doctoral Fellow, National Institutes of Health, Bethesda, Md.; William Gowen, William Prager, Vice President, Carl Byoir & Associates, New York; R. G. Hall, Asst. Director, William Markowitz, Director, Time Service Division, U.S. Naval Observatory, Washington, D.C.; Karl Hamner, Professor of Botanical Science, Harold Lyons, Brain Research

Institute, University of California, Los Angeles; F. Clark Howell, Professor of Anthropology, University of Chicago; Jacques Laporte, University of Paris; Mary Lobban, London Medical Research Council; D. J. Malan, Bernard Price Institute of Geophysical Research, University of the Witwatersrand, Johannesburg, South Africa; William Miller, Official Photographer, Allan Sandage, Astronomer, Mt. Wilson and Palomar Observatories; Philippe de Montebello, Dept. of Paintings, Bonny Young, Senior Lecturer, The Cloisters, Metropolitan Museum of Art, New York; Floriano Papi, Director, Institute of General Biology, University of Pisa; Leo Pardi, Chairman, Zoology Dept., Florence University; Colin Pittendrigh, Dean of the Graduate School, Princeton University; W. C. Resides, Allocations Engineer, National Broadcasting Co., New York; Curt Richter, Professor Emeritus of Psychobiology, Johns Hopkins University, Baltimore; Robert Stuckenrath Jr., Director, Radiocarbon Laboratory, University of Pennsylvania, Phila.; Soybean Investigations, Crop Research Division, U.S. Dept. of Agriculture, Beltsville, Md.; Samuel Sugarman, Manager, Laboratory Services, Bulova Watch Co.; Morton Sultanoff, Ballistic Research Laboratories, Aberdeen, Md.

PICTURE CREDITS

The sources for the illustrations which appear in this book are shown below. Credits for the pictures from left to right are separated by commas, from top to bottom by dashes.

Cover—Richard Bergeron from DPI.

CHAPTER 1: 8—Anthony Wolff. 10, 11—Drawings by Leslie Martin. 12—Reprinted from *Man the Tool-Maker* by Kenneth Oakley in the Phoenix Book Series (1962) by permission of the Trustees of the British Museum (Natural History). 13—Windels, Archives Photographiques. 15—Adapted from an illustration in *Psychological Time* by John Cohen © November 1964 by Scientific American Inc. All rights reserved. 17—Ted Streshinsky. 18, 19—Saudi Arabian Information Service. 20, 21—James Burke, Stephanie Dinkins from Black Star. 22, 23—Henri Dauman. 24, 25—Carl Mydans. 26—Farrell Grehan. 27—Archie Lieberman from Black Star. 28, 29—Official U.S. Navy Photo. 30, 31—Holly Callery from Nancy Palmer Photo Agency, Arthur Seller.

CHAPTER 2: 32—Farrell Grehan. 35—Drawing by John and Mary Condon. 36—Drawings by Nicholas Fasciano courtesy Botanical Laboratory of the Faculty of Agronomy and Veterinary, Buenos Aires, Argentina. 39—Drawing by Gloria Cernosia. 41—Drawing by John and Mary Condon. 43—Alfred Loeblich III. 44—Walter Sanders. 45—Don Cravens. 46—Terence Shaw from Annan Photo Features—David Lees. 47—David Lees. 48—Archie Carr. 49—Jo Conner—Fritz Goro. 50, 51—Drawing by Nicholas Fasciano, Albert C. Flores from Pix. 52, 53—Wide World Photos.

CHAPTER 3: 54—Fritz Henle from Monkmeyer Press Photo Service. 56, 57—Woodcuts by Fritz Kredel. 58—Cartoon by Robert Censoni. 59—Drawing by John and Mary Condon. 60—David Lees courtesy Biblioteca Nazionale di Napoli. 61—New York Public Library. 63—International World Calendar Association, Ottawa. 65—Bibliothèque Nationale Photo. 66, 67—Drawings by Nicholas Fasciano. Right: Donald Miller courtesy Streetar Collection, Yale Medical Library. Lower left: British Museum Photo. 68, 69—Drawings by Nicholas Fasciano—Robert Lackenbach from Black Star. 70, 71—Left: Dmitri Kessel. Right: drawings by Mana Maeda—Ferdinand Anton. 72, 73—Fernand Bourges courtesy Condé Museum, Chantilly. 74, 75—Aldo Durazzi courtesy Archivio di Stato, Siena—Bibliothèque Nationale Photo.

CHAPTER 4: 76—Giraudon. 78—Drawing by Nicholas Fasciano. 80—Tosi courtesy Biblioteca Nazionale, Florence. 81, 82—Drawings by Donald and Ann Crews. 83—Drawings by Donald and Ann Crews with Gloria Cernosia. 85—Heinz Zinram courtesy National Maritime Museum, Greenwich. 86, 87—Emmett Bright courtesy Agora Museum, Athens, Emmett Bright, drawing by Nicholas Fasciano—Walter Sanders courtesy Agyptisches Museum, West Berlin. 88—Left: Heinz Zinram courtesy Dean and Chapter of Canterbury Cathedral. Right: Heinz Zinram courtesy Science Museum, London (2). 89—Donald Miller courtesy Metropolitan Museum of Art, Dick Fund, 1957. 90, 91—Emmett Bright courtesy National Archaeological Museum, Athens, Georges Pavunich, Heinz Zinram courtesy School of African and Oriental Studies, University of London. 92—Walter Sanders courtesy Mainfrankisches Museum—drawings by Leslie Martin. 93—Heinz Zinram courtesy Dean and Chapter of Salisbury Cathedral. 94, 95—Heinz Zinram courtesy Society of Antiquaries, London—drawings by Leslie Martin. 96, 97—Heinz Zinram courtesy Science Museum. Drawings by Leslie Martin. 98, 99—Drawing by Leslie Martin—Roger Wood courtesy Ashmolean Museum, Oxford.

CHAPTER 5: 100—Lee Boltin courtesy Philadelphia Museum of Art, Louise and Walter Arensberg Collection. 102—Drawing by Nicholas Fasciano. 109—Harold E. Edgerton courtesy E.G. & G. Inc., Boston. 110, 111—Bulova Watch Company, Inc. 112, 113—Sol Mednick. 114, 115—Harold E. Edgerton. 116—Dr. D. J. Malan. 117—Ed Holbert courtesy Thompson Lightning Protection, Inc. 118, 119—Dr. Morton Sultanoff courtesy Ballistic Research Labs, Aberdeen Proving Ground—Gordon Tenney. 120, 121—Brookhaven National Laboratory, Gordon Tenney.

CHAPTER 6: 122—J.D. Lajoux extrait de *Merveilles du Tassili n'Ajjer*. Editions du Chêne, Paris. 124—Drawing by George V. Kelvin. 125—From *Outlines of Chinese Symbolism and Art Motives* by C.A.S. Williams, published by Kelly and Walsh Ltd., Shanghai, 1941. 126, 127—Courtesy of American Museum of Natural History. 129, 131—Drawings by John and Mary Condon. 133—Deutsches Archäologisches Institut, Athens. 134, 135—A. Y. Owen (2)—drawing by Leslie Martin courtesy of Carnegie Institution of Washington adapted from a drawing in *Dating the Past* by Frederick E. Zeuner, Methuen & Co., Ltd. London, 1958. 136—Istituto Archaeologico, Rome. 137—Aldo Durazzi courtesy Capitoline Museum, Rome. 138—Drawing by Leslie Martin—University of Pennsylvania Museum photo. 139—Arthur Siegel courtesy the Field Museum of Natural History, Chicago. 140—Ansel Adams from Magnum. 141—Henry Grossman. 142—Mount Wilson and Palomar Observatories photograph © California Institute of Technology—drawing by George V. Kelvin. 143—Mount Wilson and Palomar Observatories photograph by William C. Miller © 1959 by California Institute of Technology.

CHAPTER 7: 144—Ernst Haas from Magnum. 147—Drawing by John and Mary Condon. 149—From *The Michelson-Morley Experiment* by R. S. Shankland. © November 1964 by Scientific American Inc. All rights reserved—drawing by Joel Margulies adapted from an illustration in *The Michelson-Morley Experiment* by R. S. Shankland. © November 1964 by Scientific American Inc. All rights reserved. 151—Drawing by John and Mary Condon. 153—Derek Rayes photographed from *Man and Time* by J. B. Priestley © 1964 Aldus Books Limited. By permission of Aldus Books Limited and Doubleday and Company, Inc. 155 through 165—Drawings by Charles B. Slackman.

CHAPTER 8: 166—Don Uhrbrock. 168—New York Public Library. 171—Historia Photo, Bad Sachsa—Archiv für Kunst und Geschichte, Berlin. 173—Drawing by Eric Mose from *Probability* by Mark Kac. © September 1964 by Scientific American Inc. All rights reserved. 175—Drawings by John and Mary Condon adapted from drawings by J. Donovan. 177—Erich Lessing from Magnum. 178, 179—Bob Gomel, Richard Jeffrey courtesy American Museum of Natural History. 180, 181—George A. Tice. 182, 183—Wayne Miller from Magnum, George A. Tice. 184, 185—Robert Phillips. 186, 187—Reproduced from *Napoleon's Book of Fate* by permission of W. Foulsham and Co., Ltd., Slough, Bucks, England, Bob Peterson photos. 188—Ralph Crane. 189—Jules Zalon from DPI. 190, 191—Yale Joel.

APPENDIX: 193—Chart and drawings by Nicholas Fasciano adapted from a chart courtesy Science Museum, London, an official U.S. Navy photograph (top), a drawing courtesy Science Museum, London (lower left), and an illustration in *Old Clocks and Watches and Their Makers* by F. J. Britten (7th ed.), E. and F. N. Spon, Ltd., London and E. P. Dutton and Company, Inc., New York (lower right). 194, 195—Drawing by George V. Kelvin courtesy Hamilton Watch Company, Lancaster, Pennsylvania.

INDEX

Numerals in italics indicate a photograph or painting of the subject mentioned.

Printed in U.S.A.